Avionics Troubleshooting and Repair

OTHER BOOKS IN THE PRACTICAL FLYING SERIES

Aircraft Icing
Terry T. Lankford

The Pilot's Air Traffic Control Handbook—4th Edition
Paul E. Illman

Advanced Aircraft Systems
David Lombardo

Spin Management & Recovery
Michael C. Love

Piloting at Night
Lewis J. Bjork

Understanding Aeronautical Charts—2nd Edition
Terry T. Lankford

Aviator's Guide to Navigation—3rd Edition
Donald J. Clausing

Learning to Fly Helicopters
R. Randall Padfield

ABCs of Safe Flying—4th Edition
David Frazier

Be a Better Pilot: Making the Right Decisions
Paul A. Craig

The Art of Instruments Flying—3rd Edition
J. R. Williams

Avoiding Common Pilot Errors: An Air Traffic Controller's View
John Stewart

General Aviation Law—2nd Edition
Jerry A. Eichenberger

Improve Your Flying Skills: Tips from a Pro
Donald J. Clausing

Weather Patterns and Phenomena—2nd Edition
Thomas P. Turner

Cockpit Resource Management: The Private Pilot's Guide—2nd Edition
Thomas P. Turner

Weather Reports, Forecasts & Flight Planning—3rd Edition
Terry T. Lankford

Avionics Troubleshooting and Repair

Edward R. Maher

McGraw-Hill

New York Chicago San Francisco Lisbon London Madrid
Mexico City Milan New Delhi San Juan Seoul Singapore
Sydney Toronto

Library of Congress Cataloging-in-Publication Data

Maher, Edward R.
Avionics troubleshooting and repair/Edward R. Maher.
p. cm.
Includes index.
ISBN 0-07-136495-1
1. Avionics—Maintenance and repair. I. Title.
TL695 .M333 2001
629.135′028′8—dc21 2001018309

McGraw-Hill 

A Division of The ***McGraw-Hill*** *Companies*

1 2 3 4 5 6 7 8 9 0 DOC/DOC 0 7 6 5 4 3 2 1

ISBN 0-07-136495-1

The sponsoring editor for this book was Shelley Carr, the editing supervisor was Sally Glover, and the production supervisor was Pamela Pelton. It was set in Slimbach by Pine Tree Composition, Inc.

Printed and bound by R. R. Donnelley & Sons Company.

This book is printed on recycled, acid-free paper containing a minimum of 50% recycled, de-inked fiber.

This book is dedicated to Ashlee C. Maher, my granddaughter and an inspiration in the writing of this book. In addition, acknowledgment must be given to my wife Sandra Maher, who has gone through it all, and Don Hawkins of Hawkins Associates Company, Inc., who gave immeasurable support to this project.

Contents

Preface

The Internet and the news media are flooded with stories about aging wire failures in commercial passenger aircraft, numerous NASA launch delays, pilot errors, and even nuclear incidents. Why is this? The answer reveals itself virtually every week with the increasing number of commercial flights being delayed due to related smoke/wiring problems. This dilemma is now affecting the average person who flies for vacation, work, and to visit family members. The problem is getting closer and closer to home and it is becoming too obvious to ignore—and we shouldn't. So far, wire problems in GA (general aviation) aircraft haven't reached the level of commercial aviation. If GA pilots, technicians, installers, and mechanics are respectively trained fully on the planes under their care, it is less likely that GA planes will ever reveal the degree of failures evidenced in commercial planes. Knowledge is power, so it should seem obvious that it is how much we know about the nature of these problems and what direction to take that helps us to effect changes.

As a pilot or technician, you are regularly faced with wiring obstacles affecting the health of the plane. If you are a professional pilot, weekend aviator, mechanic, technician, or installer, then you have already faced many of the problems that are part of doing business in the general aviation community. Gone to the local library lately? Checked out the Internet? You won't find much on avionics; in fact, this author wrote the only book you will find that intimately covers the General Aviation avionics field. Sure, you can buy a book about avionics radios and how they work, but that isn't enough with today's aircraft electronics. Aircraft professionals and enthusiasts must use their skills and every bit of information available to ensure that their planes are safe to fly. If you can't find an answer here, simply email your question to *emaherr@worldshare.net* along with the book's ISBN, date and place of purchase, and phone number, and you will receive an answer within 36 hours or less. Please note the page and paragraph you are referencing.

As aviation professionals, every time we fly, either as a passenger or a pilot, we are putting our lives in the hands of those who apply their skills and knowledge to repairing and installing radios, landing gear, engines, and everything else necessary to get the plane off the ground and into the air. The most important thing is to get the plane, crew, and passengers safely back on the ground.

Air traffic is becoming increasingly more complex with more and more planes filling the skies. NASA is presently working to develop a safer "Highway in the Sky" for those in general aviation who fly frequently between airports. Every new device coming down the pike promising to help the average pilot get to his or her destination safely and quickly is being investigated for the benefits it may offer. How we will afford the future may very well be a deciding question, but how we maintain the equipment installed on these planes is even more important.

Knowing how avionics systems work is the first step in troubleshooting. Understanding interconnects and locating the fault is yet another step, but the final step is repairing the failed circuit or system. Because avionics systems are so complex, any help you can find is well worth it.

It doesn't matter whether you are a pilot, technician, mechanic, or installer. The contents of this book will provide direction to better comprehend the intricate complexities of how a plane and its avionics interface, and someday even possibly save your life. At the very least, it should save you a few bucks when taking your plane in for repairs. Knowing what the technician expects from you as a pilot or having a better idea why the radio has failed will surely save you the cost of this book. Avionics shops can't go wrong in having this book available to technicians and installers. There is a massive shortage of both installers and technicians and without adequate guidance, they have few avenues open for instructional information. If there is one book you should buy, this is it!

Besides providing a wealth of information not available anywhere else, the contents of this book will help you, as a pilot, to select the right radio, determine how to look for the right shop, and maintain what you buy. You bought this book to help clear away the fog and mist that surrounds avionics and their relationship to the plane and that is what it was designed to do.

Learning about avionics won't make you an electronics wizard, but it will elevate your awareness and status among your peers in the professional world. Knowing how to walk the walk and talk the talk vastly improves communication with the technician (and vice versa) and therefore makes your encounters with the avionics shop less intimidating and a lot more fun! Some of the contents of this book are intended only for the advanced avionics technician or installer, and other sections for the pilot, but when applicable simply go to the section that is appropriate for you. For the technician, understanding what the pilot has to endure may provide some understanding into their squawks and how to interpret them.

My experience in the avionics and electrical fields covers nearly forty years in aviation, from "graphite" composite structure to fabric and sheet metal. My career began in the U.S. Air Force, followed with nearly 24 years at Beech Aircraft as an avionics technician and supervisor. After moving to California, I worked for Riley Rocket and Crownair at San Diego's Montgomery field. It was there that I found myself amazed at how much clients would pay to have their avionics installed or repaired, simply because they were not familiar with the process. Too much was left up to the shop to decide, when a street-savvy aircraft owner could easily cut hundreds of dollars by simply understanding more about their plane and asking the right questions. Today, I write for several magazines, write books, and represent an avionics company. This allows me to keep in contact with pilots and avionics shops all over the world. My goals are the same, to help educate both the customer and the shops to provide the best quality maintenance at the lowest possible cost for the flying public.

As a pilot, your plane and your wallet may never have to experience the enormous cost of repair if you thoroughly understand the workings of the plane and establish a regularly scheduled maintenance program. This recommendation is doubly important for the maintenance people who do the work on their customers' planes. The technicians and installers must be extremely alert, versatile, and thoroughly familiar with the avionics in their charge if they want to do the best job for the least cost to the customer.

Take the time to learn all you can about the aircraft. Avionics systems depend on a flying platform that hopefully won't spring any leaks, but like my second supervisor at Beech Aircraft Corporation once said, "There aren't any guarantees, just opportunities. The trick is being ready to take advantage of them when they come along." Hopefully, that is what this book will do for you. It will provide the knowledge and therefore the opportunity to make the technician's job easier and the customer's flying experience as pleasurable as he or she wishes it to be. Success is in your hands!

Each chapter in this book covers one avionics system or technical process. After a brief description of the system, installation tips are listed for that particular system. These tips cover specifics that must be discussed between the avionics installer, the technician, and the pilot to make sure the job is done to the highest standards and to their mutual satisfaction. A section follows this on basic troubleshooting, which details the problems you're likely to encounter.

If you wish to solve a particular problem, you can jump directly to the respective chapter and learn more about how to identify the problem and even do some basic troubleshooting. Remember, for pilots it is very important to keep detailed notes in your squawk book and set aside some time to discuss the problem with your technician. The technician will have that wealth of information to draw upon if the pilot's squawk (failure) notes are kept legible and detailed. If the problem is intermittent or complex, the pilot may have to try to duplicate the conditions under which the failure occurred. If cost is less of a problem than the inconvenience of having the planes grounded for repairs, there are facilities that have the expertise and equipment to find that illusive intermittent (open or short), but be advised, it will cost. Test equipment for this level of troubleshooting may cost over $50,000 for one test system. This is a lot to pay for test equipment and that cost must obviously be passed on to the client. This is why the participants must have as much information on hand as possible. Who knows—as a pilot, you may find the reason for a given failure before your wallet hits the counter. If you are a technician, you will have saved your client some big bucks by repairing the failure quickly. As a kit builder, there is nothing more frustrating than having to fix your plane before you can lift off for the first time. If you aren't faced with a problem yet, just sit down, enjoy this book, and learn as much as you can.

About the Author

Edward R. Maher is the West Coast VP for Hawkins Associates Company, Inc., an avionics distributor with over 2,500 dealers across the United States and parts of Europe. He began his career as an electrical mechanic in the Air Force, then worked as an avionics technician and manager for Beech Aircraft Corporation, then later managed the Avionics shop at Crownair. Mr. Maher is the author of three other books on aircraft and is a contributing writer to *Avionics News*, *Kitplanes*, and *Aviation Maintenance* magazines.

CHAPTER 1

Rules to Fly By

Scheduled Requirements

The aircraft unceasingly depends on a collage of complex, interacting systems to get the pilot and passengers safely from one point to another, but if repetitive maintenance isn't performed as required, a critical system could fail, endangering the plane. To reduce the possibility of this occurring, regularly scheduled inspections are mandated by the FAA. This includes, but is not limited to Title 14 of the Code of Federal Regulations which includes the transponder, altimeter, plumbing, Emergency Locator Transmitter (ELT), etc. As important as scheduled inspections are for your automobile, they are triply important for *any* aircraft. We have seen what happens when inspections are ignored, overlooked, or deliberately not done and the results are not pretty. Everything from control surfaces to altitude failures is attributed to sloppy maintenance. Doing it right the first time may take longer, but earns frequent flyer points over the long haul!

1. Your plane should have an altimeter/static system test, even if the plane isn't flown IFR (Instrument Flight Regulations). The plane will still show up on radar, and it certainly helps to know where you are. This is the most efficient way to keep the static

system and altimeter accurate, and it only costs about $50 every two years, depending on individual shop pricing. Certification should not be a matter of cost; this is a statement of security, safety and reliablity. Reference Title 14 of the Code of Federal Regulations (14 CFR) part 91, section 91.411. See Chapter 16 for more detail.

2. The transponder and encoder must be tested every two years per FAR 91.413. This test will check the transponder output and the accuracy for the encoding altimeter. Cost for this is between $100 to $175 every two years. Review Chapter 16, Reference Title 14 of the Code of Federal Regulations (14 CFR) part 91, section 91.413.
3. The ELT battery must be replaced at intervals marked on each cell, and all cells must have the same date. Additionally, the date must be displayed on the outside of the ELT case. Replacement of the ELT cells may be performed by the pilot if the process can be defined as "simple" (like changing flashlight batteries). Otherwise, they must be changed by a qualified shop. Should the old battery leak or display other forms of deterioration it stands a good chance of being discovered during the scheduled replacement period, thereby preventing further damage. Lithium batteries have an eight-year replacement window under normal operating conditions; however, should the ELT be operated for an unknown period, the battery must be replaced. Alkaline batteries must be replaced every four years and again, if the ELT is operated for an unknown period, the batteries must be replaced. Later, this book discusses how to test ELTs so you won't be caught with a dead unit when it is most needed. Title 14 of the Code of Federal Regulations (14 CFR) part 91, section 91.207 (more on ELT in Chapter 17).
4. A VOR check is due every thirty days. The easiest way to do this is to fly over a checkpoint near your home base. Repeat the test on a monthly basis at the same checkpoint to assure that the system is still in compliance. If there is a sudden change in the test

results, confirm with another checkpoint. There may be some technical difficulty that has not been reported. It is a good idea for pilots to log the test results in a separate avionics log where they keep track of all avionics work done on the airplane.

5. The aircraft radio station license is due for renewal every five years. It is an easy one to forget, but the record keeping is necessary, and the certification must be done!
6. Vacuum filters. There are usually two filters in most light aircraft: a central vacuum filter that needs to be changed every 500 hours, and a vacuum regulator filter that should be changed every 100 hours.

How do you keep track of all these events? With so much going on in your private and business life, it isn't hard to miss one or more of the requirements: the annual inspection, the static/transponder test, the ELT battery test, the VOR check, perhaps some recurring airworthiness directives and service bulletins, radio station license, vacuum filters, your medical certificate, biennial flight review, and your IFR currency. A lot to remember? Yes, definitely, but with some form of dedicated planning system it can and *must* be done.

To ease your search for an Aviation Electronics Association (AEA) facility, the AEA puts out a Pilot's Guide that lists every member. This guide provides locations of certified avionics and instrument facilities, manufacturers, distributors, and even academic institutions around the world. [Call (816) 373-6565 for your copy and to order your free subscription to *Avionics News*.]

The simplest way to keep track is to use a day planner, computer alert, or a bulletin board. If you're not already in that habit, you can pick up a simple calendar/planner at an office supply store and use it exclusively for your aviation tracking. Computer software programs can be found as shareware on the 'net for both the Mac and PCs. If you are into high-tech gadgets, look at the Palm Pilot or Visor handheld computer. Both use the same operating system and will work quite well to download to and from Macintosh and PC compatibles.

Whenever you complete an event, make a note of when the next one is due. It's that simple. For instance, when a pilot has a static/transponder check done on his or her new plane, say in August 2001, the pilot would make a note in the planner that the plane's static/transponder check is due August 2003. The same can be applied to the ELT battery. When you install a new one, note the replacement date in your database or planner and you'll never forget it. Another useful idea is to note when all these items are due next to the signoff for the annual inspection. That will ensure that you don't miss any items that are coming up soon. There is also one more easy way to create a reminder. If you have voice mail with a wake-up call or reminder feature, take advantage of it. You are most likely already paying for it, and it is easy to use.

Don't try to rely on memory for these important events. It's too easy to forget to write it down, then one day—surprise—it's been a year since your biennial flight review was due. If you're using a computer program to keep track of your flight hours, you can probably plug in the due dates for the static/transponder test, ELT battery, VOR check, and other recurring maintenance events.

Don't forget FAR 91.403 (a): "The *owner* or *operator* of an aircraft is *primarily* responsible for maintaining the aircraft in an airworthy condition. . . ."

If a seller sells an airplane that isn't airworthy because the altimeter is inaccurate, would it be the responsibility of the mechanic to make sure the airplane was airworthy? Was it the mechanic who completed the repurchase inspection? No, in the final analysis, it was the owner's responsibility, and the owner's noncompliance with the regulation can cost the pilot and crew their lives.

How to Squawk a Problem

Remember science and math classes in high school? The teacher drummed into your brain that more than half the solution to a problem lay in defining the problem first. It is no different when solving difficult and complex avionics problems. Know how the systems work, understand their operation in varying environmental conditions, gather information, know the right questions, and listen.

The goal for you as a pilot in squawking a problem is to give the avionics technician enough information so the tech can zero in on the problem and begin working to repairing it, not spending your hard-earned money doing what you could easily have done. Keep a notebook handy whenever flying so you can make notes that will speed along the troubleshooting process. The same is true for the technician assigned to solve the squawk. The technician should keep his or her own database of squawks and repairs. If a pattern begins to appear, it could be the key that solves the problem.

Step one is to isolate the problem. Is it intermittent or does it happen all the time? Are the circumstances the same each time it happens? If there are two of the malfunctioning items, is the other one working normally? Are you sure you are using the equipment properly? Double-check the operating manual and pilot's guide. Note the conditions of flight during which the problem appeared, including altitude, position, attitude atmospheric conditions, other avionics status (on or off), other electrical equipment status, or whether engine speed or retractable landing gear operation affects the problem.

Second, spend some time checking for obvious defects such as broken wires, loose connectors, poor bonding (especially at antennas), excess harness stress on connectors at the back of the radio stack, or moisture leaking onto radios. If it is a COMM radio problem, have you tried the spare microphone? Failing to do this is very common with those who do not fly professionally. Professional pilots, backed with extensive training and experience, will always check the microphone. One recurring thing that can cause microphones to fail is broken wires directly where the wires exit the microphone case and at the microphone plug.

Your input to the technician is extremely valuable and will go a long way toward solving complex problems. If the technician insists he or she can't duplicate the squawked problem, the pilot should consider taking the technician for a test flight so the pilot can demonstrate the problem firsthand. It would seem to be obvious that the pilot would want to make a sure he or she can duplicate the problem for the technician's benefit. This isn't always the case, but should be. The pilot is the critical link in resolving any aircraft failure and should be part of the solution, not the problem.

Keeping Up with Modifications, Bulletins, and Airworthiness Directives

Like any part of the airplane, the avionics may have Airworthiness Directives (ADs) and manufacturer's service information or service bulletins and letters that haven't been addressed.

An AD will reference serial numbers of the equipment to which it applies. If the aircraft operator/owner gets an AD in the mail, he or she should check the serial number applicability, and if it doesn't apply, make a note on the AD and put it in the aircraft records.

There probably won't be much mentioned about the manufacturer's service information unless you contact the manufacturer and ask to be put on the mailing list for new bulletins. Usually, there will be an annual subscription fee; however, your avionics shop should be receiving bulletins for your avionics and letters from the manufacturer. It is up to the shop manager to notify the pilot or technician, should a bulletin apply to one of the aircraft's radios. Note, if the shop is doing its job right, it provides them an excellent opportunity to sell additional services or new toys. This is why aircraft owners should seriously consider having only one shop where they get the bulk of their service work done.

In some cases, the suggested modifications are not important, but in other cases, a bulletin may be directly applicable to a problem the pilot is having. When an airplane is brought in for service, the shop manager should let the plane's owner know if there are service bulletins or letters applicable to the plane's equipment that should be taken care of or that cover an existing problem. Keeping the plane's avionics up to date with the latest modifications and upgrades will make the plane last longer and increase its long-term value. Don't ignore manufacturers' service information.

It is highly recommended that the aircraft owner contacts the manufacturers of the radios rather than wait for them to contact the shop. Some manufacturers are slow to release the updates or are waiting for the radios to come in for repair. So, don't wait—check on them yourself; some of these updates are relatively inexpensive and provide improved operation.

Keep in mind the ADs are mandatory, and therefore compliance is obligatory. Manufacturers' service information is always optional, no matter what the manufacturer tells you, unless specified as being required for compliance with an AD.

One thing both pilots and technicians can do to keep abreast of what is happening in the general aviation industry surrounding avionics is to subscribe to *Avionics News*. There is no cost to you and the information can be invaluable. Call (816) 373-6565 to order your free subscription.

CHAPTER 2

Avionics Shop Requirements

Technician Certification

It is not the intent of this book to teach you how to delve into the innards of your airplane's black boxes (avionics) and solder a new transistor on a deeply buried circuit board. Sure, some soldering will be covered to provide a better understanding of what you should expect from the shops and the manufacturer, but unless you have forty hours or more of training, don't touch the circuit! That is the minimum level of training the people who build the harnesses and circuit boards receive at the aircraft factory before they even touch a real harness or radio. The same is true for crimping and tool use. The information is there if you need it, but as mentioned earlier, apply the knowledge respective to your corresponding skills and experience.

Aside from the hands-on descriptions and photos, pilots and technicians will get the most from this book if they use the knowledge to help isolate and troubleshoot problems to get them fixed quicker and less expensively. The bottom line is that the pilot and shop will have to function as a team, to mutually develop a good working relationship between the avionics technicians and installers. Speaking of which, who are these mysterious people?

Most technicians have spent long years training and retraining, just for the privilege of getting their hands on those expensive black boxes. However, for anyone to work on *any* kind of transmitter, the Federal Communications Commission (FCC) must issue a license to a technician, after he or she successfully passes a test. While it is possible to simply learn what is necessary to pass the test, most qualified technicians have already secured a degree in electronics or extensive training in electronic repair and maintenance. The FCC license is usually integrated into the program with testing at the end of the school year. One good way of getting this experience, for instance, is in the military. Many young men and women are entering the military to get training as part of their long-term goal to develop a career they normally couldn't afford. While part of the armed forces, it would be best if they have their FCC license prior to terminating service.

You'll frequently find that technicians working for avionics shops are split into a few specialty areas. Bench technicians are the people who dig into the black boxes after they've been removed from the aircraft. Bench technicians also specialize on specific radios. One will work mostly on pulse-type transmitters [DME (Distance Measuring Equipment) or XPDRs (transponders)], while another will be the shop's autopilot authority or resident expert. Other technicians might specialize in areas like installation, troubleshooting, and customer service. Some large shops, for example, have a person assigned to work with customers on troubleshooting avionics, sometimes flying with the customer to isolate problems before avionics are removed or repaired. (Wouldn't you like to eliminate this step?) It's a lot more efficient to troubleshoot first than to waste a $70-per-hour technician's time opening up a black box that isn't broken. Also, as you'll learn, there are many errors caused during installation and must be examined with the radio installed, looking at the system as a whole.

Bench technicians must have a FCC certification to work on transmitters like COMM (communication) or pulse equipment radios. For all other radios (receivers only), such as marker beacons, ADF, etc., technicians must be licensed as an FAA (Federal Aviation Administration) airframe mechanic. This means they will carry an FAA Repairman's Certificate covering the specific type of equipment he is approved to work on, or simply work for an FAA-approved repair station. There is not yet a formal FAA certification for avionics technicians, although many airframe and powerplant (A&P) schools do offer separate avionics courses leading to FCC (Federal Communications Commission) certification and usually include FAA Airframe Mechanic certification.

Avionics Shop Certification

Most avionics shops are FAA-approved repair stations. An owner of a shop has two choices when it comes to legal status. The owner can elect to either go the non–FAA-approved route, or seek approval, which provides a degree of credibility to those seeking maintenance.

Lack of FAA approval simply means the shop isn't an official repair station. To operate with repair station status, the shop must apply for and be approved as a repair station by the local FSDO (FAA Flight Standard District Office). This involves showing proof of meeting certain requirements such as experience of management, ability to install and repair various types of avionics, commitment to stocking spare parts and necessary test and repair equipment (including calibrated tools). Other requirements might include providing a clean room for instrument repairs or routine training for technicians.

Many avionics shops seek official repair station status because it ensures customers that the shop meets a minimum verifiable set of FAA standards; in addition, it allows the shop to hire nonlicensed individuals to do the actual work. There is no exception to the requirements for an FCC-licensed tech to work on transmitters, however a repair station can technically hire anyone off the street and put him or her to work performing whatever other work the repair station is authorized to do. This does not mean that a non-licensed installer working for a repair station is less qualified than a FAA-licensed airframe mechanic working for a nonapproved avionics shop. In fact, because of the stringent requirements attached to repair stations, the people who work there usually receive considerable training and are very well qualified for the work they do.

Many repair-stations like to hire FAA A&Ps (airframe and powerplant mechanics) because of the excellent training they receive.

Although not required, most shops prefer to employ FCC-licensed individuals for bench work, even if the FAA or FCC doesn't require them to do so (a FCC license is required when working on transmitters). There is a sense of pride and, yes, responsibility that is part of being FCC-licensed. The quality specification for work performed on land-based electronics is not equal to FAA requirements. Avionics work must be performed to specifications that are more stringent. After all, you can't pull over and park the plane if you run into problems mid-flight.

Technically, a licensed airframe mechanic is the only person allowed to remove and install a radio unless the shop is a repair station. Yes, this means simply removing a radio from an instrument panel, which usually requires a twist of a screw. The aircraft owner is not technically allowed to remove the radio, but remember who is sitting in a chunk of metal flying at eight thousand feet with no radios. The radios might simply be loose in the tray, or suffer from corrosion or a combination of a bad radio and corrosion. Many pilots have revived their radio by simply switching the number 1 and 2 (most planes are equipped with two similar radios for redundancy). I want to be clear: I don't want to give permission for the pilot to be removing the radios, but remember the final responsibility belongs to the aircraft pilot.

For a non–FAA-approved shop, each person who works there must be appropriately licensed. The airframe mechanic removes and installs avionics for the bench tech to repair; bench techs who are not airframe mechanics must perform repairs under the supervision of one who is. The bench tech gives the repaired radio back to the airframe mechanic for reinstallation. The airframe mechanic would have to perform all the harness building, wiring, and physical installation work for a new installation. A licensed airframe and powerplant mechanic holding FAA Inspection Authorization (IA) would have to sign off the FAA form 337 to make the installation legal.

This would obviously be an inefficient way to run an avionics shop. You'll find shops that do operate like this, where the bench tech removes and installs the radios being worked on, and nonlicensed personnel assist on installation. Technically this is illegal, but it does happen. I mention this not to sic the FAA on a bunch of rule-breakers, but to let you know that some shops might not comply with the greater percentage of the rules. Typically, the more a shop complies with the rules and shows a willingness to work with the FAA, the better the quality of the shop's work.

Repair station status confers certain benefits on avionics shops. While it does involve more surveillance by FAA inspectors and more stringent record keeping, earning a repair station certificate more than makes up for the increased paperwork by allowing improved efficiency. To obtain a repair station certificate, the owner of an avionics shop must meet certain minimum requirements. These include:

1. Hiring personnel with a minimum specified amount of experience.
2. Setting up procedures for handling repairs, processing parts, and maintaining test equipment.
3. Recurring employee training.

When a company has a repair station certificate, it need not hire licensed personnel to perform the work. Standards are set by the FAA, and the heads of the various departments in the shop are responsible for enforcing those standards. A repair station, for instance, can hire an unlicensed avionics installer. Just because this person does not have an airframe mechanic license does not mean he or she is not qualified to do the work. There are many installers, for instance, who formerly worked for airframe manufactures and never obtained their airframe license; they would be ideal candidates for installation work at avionics shops. It is interesting to note that even with the flexibility that a shop has, there is a severe shortage of avionics installers and technicians. Pay for avionics and installer positions are far greater now than just five years ago. Even installers are being paid as much as $18.00 per hour, as opposed to $10.00 in 1995. In some shops, a good installer can get as much as $21.00 per hour.

Technicians and Management Responsibility

The head of the installation department oversees all the work done in that department and ensures that it meets company and FAA quality standards. The FAA requires that each repair station employ an experienced inspector to maintain quality levels. Essentially, the experienced inspector and department heads guide the unlicensed personnel.

Another benefit is that the repair station transfers the burden of paperwork from the person who did the job to the manager. At a nonapproved shop, each licensed person who did the work would be required to fill out logbooks, Form 337s, plus weight and balance updates.

The repair station management is responsible for all the paperwork, leaving the technical personnel free to install and fix radios and keep the revenue stream flowing in the right direction. This points out another drawback to the nonapproved shop. In reality, many mechanics hate paperwork and gladly leave it for managers to fill out. Technically, this is not legal. The person who does the work must fill out and sign the logbooks, but in the field, you'll find this isn't always the case. Not only is it more appropriate for the mechanics and technicians to complete the paperwork, it presents less of a challenge for the supervisor to "create" the closing forms and documents.

This long explanation is simply an introduction to shops and what pilots and job-seeking technicians or installers have to expect. Chapter 3 of this book includes tips for pilots to deal effectively with their avionics shop and what to look for when evaluating a new one. Also included is what pilots need to look for in terms of the paperwork that must be done whenever they have radios repaired or installed.

Work Pilots/Owners Are Allowed to Perform

Unfortunately, or fortunately, depending on your stance, when it comes to avionics, the FAA has pretty much left aircraft owners with little they can legally do to repair their own radios, unless the pilot is the manufacturer (example, kit planes).

FAR part 43, Appendix A details the categories of maintenance under which various repairs and alterations fall. There are two types of repair or alterations: major and minor. Major repairs and alterations must be signed off by an IA or by an authorized repair station (meaning that the repair station's certificate permits that specific repair or alteration). An airframe and powerplant mechanic can do minor repairs and alterations.

As the owner of the aircraft, you are allowed to perform preventive maintenance. Items that are considered preventive maintenance are listed at the end of Appendix A of FAR Part 43. That list doesn't include one item of avionics-type wiring repairs: In fact, the only mention of wiring at all is that you are permitted to repair landing light wiring.

So Where Does that Leave the Aircraft Owner or Pilot?

There is one legal way you can work on your aircraft. You can work on your aircraft as long as you are supervised by an appropriately licensed person, such as an AP. This person must directly supervise your work and sign it off in the logbook.

An airframe mechanic is permitted to perform minor repairs on aircraft and their related appliances. This includes the avionics and all the wiring. Outside of an FAA-approved shop, an airframe mechanic is not permitted to perform major repairs on an appliance like a navigation receiver. An example of a major repair is calibration of a VOR (Vector, Omni, Radial) indicator. This must be done either by an IA or by an approved repair station. Basically, an airframe mechanic is limited to removing and installing avionics, repairing wiring, and inspecting of installation for 100-hour inspections. A mechanic with the appropriate FCC license could repair transmitters as well.

It is important to realize that pilots are not allowed to remove their own #1 NAV/COMM (Navigation/Communications) radio. In other words, they cannot remove any avionics component and bring it to an avionics shop for repair. It may sound ridiculous, but those are the rules, and the pilot/owner should be aware of them. Any time a mechanic removes a transmitter for repair, the mechanic is required to log the removal and the subsequent reinstallation in the aircraft's logbooks. Hardly anyone does this, but FAR Part 43.9 mandates logging all maintenance, regardless of whether it's a major or minor repair or alteration, or preventive maintenance performed by the owner. Chapter 3 explains this in more detail, as well as why record keeping is so important.

To recap, if you are the pilot and want to work on your own avionics equipment, perhaps repair the broken wire on the push-to-talk switch or resolder the wire that is causing an intermittent problem on your microphone jack, you need to do it under the supervision of at least an airframe mechanic. After the repair is completed, an A&P mechanic must sign it off in the logbooks. You'll also need a mechanic to supervise and sign off your removal and/or reinstallation of any piece of avionics.

Field Approval (Form 337) is a tool used by shops to install or alter the aircraft. FAR, Part 21.305 covers the approval of components and appliances by ". . . any other means deemed appropriate to the Administrator." This is for "parts-related" approval only. The actual aircraft alteration approval is not governed by any regulation; however, FAR Parts 43.3 and 43.7 come close. The management and submittal of a Form 337 is regulated in FAR Part 43, Appendix B. Any Aviation Safety Inspector (ASI) who has attended the appropriate courses can sign off any modification or installation without any other authority and, in turn, take full responsibility for it. This is important; not just *any* ASI can sign off a modification or installation without any other authority.

ASIs must have first attended the FAA Alterations Course. For avionics, they must also attend the FAA Avionics Certification Course; then they are authorized to approve any alteration that is deemed to be a Major Alteration (Minor Design Change to the type-certification basis). The ASI is specifically not authorized to approve, for example, TCAS II, EFIS, DFDR, etc. They can require that they personally inspect the modification or installation or sign it off based on data alone. Of course, this would require an inspector who has considerable experience and the confidence to inspect and approve the installation. If a field inspector is not technically "street smart" in the area in which he or she is willing to approve a 337 installation, the Field Approval process could be compromised. Besides the courses, a degree of experience must be applied to any inspection.

There is no question that the STC (supplemental type certification) process is far superior to the Field Approval process, however, this assumes that the ACO is fully qualified to process the STC. The basis of certification is established in this process and objective evidence is prepared to support the claims for certification. The FAA is responsible for safety assessment throughout the STC process. It is regulatory bound, and it is a form of attesting to the certitude of a product (aircraft, engine, or propeller). The field approval process falls far short of fulfilling those expectations and requirements, but it does have a place in the approval process. If not for the Form 337, the Loran and GPS would, most likely, still be touring in boats. In the STC process, the applicant makes a "showing of compliance" and the FAA makes a "finding of compliance." The Field Approval process recognizes only a "declaration of compliance" and a FAA "acceptance of compliance."

The FAA states,

> Acceptable data (data that is acceptable to the (FAA) Administrator) is comprised of such information as installation manuals provided by avionics manufacturers, AC 43.13-1B and AC43.13-2A, service bulletins, and avionics maintenance manuals when they are referenced in an installation in an aircraft (product), however they are approved when repairing an appliance, e.g., radio, intercom, autopilot, etc.

The STC was indirectly originated by Jack Riley, Sr., back in the 1950s after the CAA (Civil Aviation Administration) felt that a Form 337 wouldn't cut it any longer after he converted fifteen single-engine Navions to what was to be known as Twin Navions. They felt that the modifications were a little more than minor alterations and another method was necessary to accomplish a level of certitude necessary to ensure safety and public confidence.

There are two types of STCs; one is a multiple STC, where one can cover several aircraft models, and all serial numbers of those models, except those in which a certain modification may be specifically prohibited as identified within the Type Certificate Data Sheet for the aircraft. The second type is a "one aircraft only" STC and may not be used on any other make, model, or serial-numbered aircraft. It is specifically limited to "one aircraft," just as is an approval using a Form 337. The single exception of the Form 337, Field Approval, is that if *identical* alterations can be accomplished using the *same* approved data on one Form 337, such as with the 62 Bell 206s into which Petroleum Helicopter installed VFR-limited GPS equipment, using the same approved data. Representatives within the FAA have specifically stated that they wouldn't have a problem, if an STC were approved, to use the data to install the same equipment into different aircraft using a Form 337.

The TSO (Technical Specification Order) is theoretically an honor system. The FAA merely examines the technical data submitted by the manufacturer for continuity and completeness. The FAA doesn't analyze the design. Remember, a TSO is only issued to Class II products, meaning appliances and components, not to aircraft, aircraft engines, or propellers, which are Class I products. Any "post-installation" requirements are *not* a part of the TSO.

TSO documents specifically recite:

> The TSO identifies the minimum performance standards, tests, and other conditions applicable for issuance of design and production approval of the article. The TSO does not specifically identify acceptable conditions for installation of the article. The TSO applicant is responsible for documenting all limitations and conditions suitable for installation of the article. An applicant requesting approval for installation of the article within a specific type or class of product is responsible for determining environmental and functional compatibility.

The TSO process is covered by regulation with the basis of approval for an installed TSO article being independently considered. FAR 21.605 through 21.617 deals only with processes for applying for quality assurance measures for, modifications of, and letters of design approval for articles intended to be TSO'd. If an STC is warranted, then the IA or other authorized person such as a certificated repair station may "sign off" the return-to-service entry, Block 7, even after the STC is obtained, which must be identified on a Form 337.

The FAA is not requiring STCs for equipment that is not major, unless the equipment, which is being installed for the first time, is new, novel, or unusual or if it has to be approved by STC by policy. Examples include special conditions issuance for high-intensity radiated field (HIRF) or lightning "validation," or for equipment which causes a major change to the type design, such as Traffic Collision Alert System (TCASII), Digital Flight Data Recorder (DFDR), or Electronic Flight Instrument System (EFIS). All require systems integration analysis and testing. Equipment that is "highly integrated," or equipment that is introduced into an aircraft for the first time should undergo a formal FAA evaluation for functionality, readability, performance "verification," etc. This process can only be accomplished by an STC.

The problem when translating the FAA's stance on this is, when is it considered new, novel, or unusual? If the new system is not integrated or impacts the intent of the installed avionics, why would it require an STC? These are questions that are flying around without a perch to land on.

Note that the TSO denotes the performance standards under standard and environmental conditions and the required documentation, but doesn't say anything about the quality of the equipment or the components used. Only the quality of the manufacturing process is checked, but not necessarily by qualified (manufacturing professionals) people.

Because the Radio Technical Commission for Aeronautics (RTCA) or the Society of American Engineers (SAE) usually generated the standards for new TSOs, this became a lengthy process. This created one problem. Another problem is that the local ACOs were required to generate the documentation that would be the basis for the new TSO.

Personnel in the local ACOs (Aircraft Certification Offices) were either not competent or unwilling to expend the effort to do this, so they promoted piggybacking on existing similar TSOs. This led to abuse as a result of too-liberal interpretation of existing TSOs, so the ACOs (and Washington) have decided to discourage this process altogether and push the PMA (part manufacture approval) process for avionics.

To discourage the TSO process, the ACOs could extend the certification process indefinitely by rejecting legitimate proposals for similar applications by the manufacturer. (It's as if they are saying, "Keep submitting piggybacks to different TSOs until you submit one we like. Don't ask us in advance which one we like or we aren't going to like any of them.") The only solution for manufacturers is to write their own TSO and submit it to the ACO in their district. This will minimize the effort and required competence of the ACO personnel and promote their support.

The PMA is covered by regulation. The PMA allows a company to manufacture a product that has been type certificated (Type Certificate or Multiple STC). It also authorizes a company to manufacture another company's product that has been TSO'd. Document and manufacturing capability evaluations are similar to that for a TSO. The difference is that the design and manufacturing documents and results of tests of typical completed parts have to be submitted for review, evaluation, and approval.

The PMA does not authorize one company to manufacture another company's product that has been TSO'd. TSO holders may, in fact, choose to have some or all of their articles manufactured "out of house," and there happen to be manufacturers who hold PMA who produce circuit boards and other subassemblies for TSO holders. UPS Aviation Technologies is one of those companies. The documentation preparation and submittal requirements are generally the same for PMA'd and TSO'd articles. There are usually two specific reasons why a PMA is obtained for avionics equipment that are intended to be installed in the system as TSO'd articles: A TSO doesn't exist (yet) for that particular avionics equipment, or the article can't meet the requirements of the TSO, such as GPS navigation equipment containing a GPS engine that does not have receiver autonomous integrity monitoring (RAIM). The requirements are the same for all documents submitted for "review, evaluation, and approval."

Parts such as capacitors, resistors, connectors, etc., are not TSO'd or PMA'd. They may be manufactured under some military or industry standard that is acceptable to the FAA or just identified as parts that conform to certain industrial standards. FAR Section 21.303(b)(4) covers the description and standards for "standard parts." They typically include nuts, bolts, and other fasteners manufactured under MIL-Spec, AN, or another U.S. standard. The same is true based upon the work that the AEA (Aviation Electronics Association) did to promote electrical and electronic discrete parts, other than programmable logic devices, microprocessors, etc., including connectors that are not standard parts. Connectors, other than those specified in MIL-Specs must be PMA'd or accepted into a TSO design. Many are not, and they are used in installations wherein the avionics manufacturer does not supply them with the TSO'd or PMA'd article as part of the installation "kit."

An STC is for one aircraft, by serial number, or for multiple aircraft by model number or by type certificate. For example, a "multiple" STC may be applied for and considered in the drawings and other technical data developed to support "multiple" configurations for installation of equipment. This would include several different variants to take into account differences in equipment locations, cable routing, etc. Such configurations may extend to Cessna 172, 182, and 206; for example, all aircraft that were awarded the same type certificate.

PMA may be obtained only by applying for and demonstrating parts or material conformity under a "multiple" STC. A TSO does not consider the aircraft make, model, or type certification basis into which such a TSO'd article is to be installed. The exception is wheels, brakes, and other aircraft-specific equipment and components, which are listed in the TSO authorization letter.

Avionics equipment is typically not limited in such a manner, but there are exceptions, such as with GPS/WAAS navigation equipment intended to be installed discriminately within FAA Part 23 airplanes under 6,000 lbs., by class versus those for installation into commuter-class airplanes. This is described within RTCA/DO-229 and the TSO.

CHAPTER 3

Avionics Decisions

Choosing Avionics

The decision by the aircraft owner to buy a new avionics radio is usually made because of repetitive failures, such as an intermittent problem with no failure found (NFF), which thus never gets fixed, a noisy radio, drastic FAA changes, or simply a desire to enter the modern age. Avionics technicians will try to recommend replacement of these older radios simply because they know the customer is throwing their money away. Over 80 percent of aircraft built since the late 1960s have not incurred any major radio modifications. If you end up with one of these airplanes, you'll no doubt see at least one ancient radio lurking in a well-worn instrument panel. Many of these old radios are tube-type and so tired they are ready to take a long vacation in a dark cabinet, maybe a museum. While it may be possible to get these radios working, you have to wonder how long they'll continue to operate and will they do so reliably? Ask yourself: Will this old radio be a bottomless pit into which I will be constantly throwing money? The answer has to be, "Yes." There is no other answer. There is no way that older radios can accurately maintain the same level of efficiency expected of the modern radio. Everything from heat to cold affects their stability, not to mention simple component degradation and failure.

Avionics Radio Retrofits

One manufacture, McCoy Avionics, offers a retrofit front-end for the King KX170-series NAV/COMM. The MAC 1700 series Control/ Display unit replaces the usually worn-out mechanical frequency selectors in the King radio with a modern, digital-readout, multiple-memory front end. (The upgrade kit costs between $1295 to $1495, plus labor for installation, which typically runs about $300). One problem to watch for, however, is the condition of the rest of the radio, the rear end. What's the sense of spending that much money to upgrade a radio that is on its last legs?

If the customer decides to go this route, the technician should recommend that the customer have the old NAV/COMM bench-checked prior to any modifications to determine if its components are operating at their proper levels. Also, investigate for any upgrades and modifications that are available and should be performed as part of the front-end replacement. This is to ensure the radio will last for a reasonable amount of time. Another low-cost approach is simply removing the old radio and replacing it with a new direct-replacement radio that slides into the existing rack. Narco offers these slide-in re-

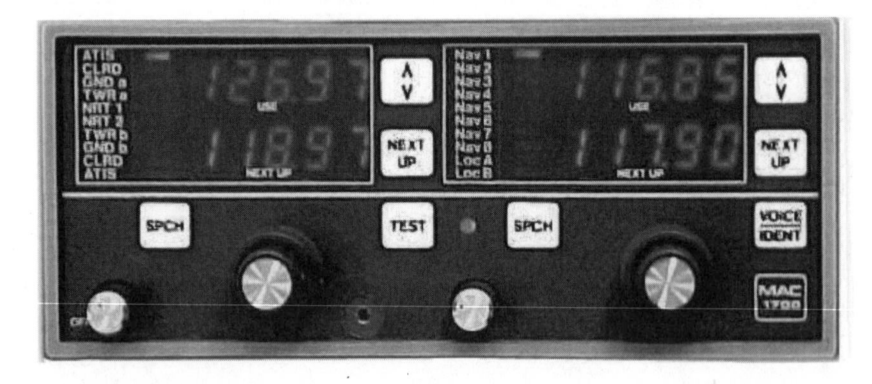

TKM MAC 1700 NAV/COMM.

placements for older Cessna-made ARC radios. TKM (Michel Aviation Productions) makes slide-in replacements for older King, Narco, and ARC radios. In many cases, no modification is necessary; just slide the new radio in and away you go. Even some of the older tube-type ARC radios can be replaced this way, although in some cases an adapter kit is necessary (short adapter harness). This is a good way to upgrade the NAV/COMM, except for the wiring, which can be a weak link in the whole system. The wiring, by itself, could be a source of noise and annoying intermittent problems that weren't so evident with the older radios because they were less sensitive than the new one.

Narco NAV/COMM.

If the pilot suspects any kind of wiring problems, have them taken care of before spending money on new, slide-in replacement radios. A few words of advice: Corrosion and other contaminants can and will coat the surface of the contacts. Clean the mating contacts in the connector before installing the new radio. First, place a white cloth on the bottom of the tray to catch any residue as it falls. Using a cotton swab dipped in alcohol, wipe the contacts using an in and out motion. Sometimes, a white ink eraser is needed to remove difficult materials like cigarette tars. The reason? Surface-to-surface contact between the new radio and the old connector will not be the same, and now the nonconductive contaminants on the contacts may prevent normal operation or no operation at all. The connector contacts may need to be reformed, but that is dangerous territory and should only be accomplished by a skilled technician. (The contact fingers would need to be lifted slightly using a dentist's awl or probe, without damaging the connector. One end of the contact [open end of connector] is held with gentle pressure by the ink eraser, while the probe is doing its job of lifting.)

Cutting corners with avionics communications systems is like setting a time bomb in the instrument panel. There isn't any way you

can win or know when electronic components will fail, however, the odds of failure are greater when the radio already has two strikes against it. Using old radios with a new front end may help you operate the radio without mechanical frequency selection problems, but doesn't eliminate the possibility of electronic failure in the rest of the radio. Don't forget the wiring; relying on existing wiring that may be over twenty years old is further stacking the deck against safety and reliability.

If the plane has one of the following listed radios, it can be replaced with a fully modern, slide-in unit offered by either Narco Avionics, or TKM Michel, or the front end replacement by McCoy Avionics.

TKM radios can replace the following ARC radios (Aircraft Radio Corp, formerly owned by Cessna Aircraft Company, since sold to Sigma-Tek):

- RT308B
- RT328A
- RT329C
- RT328D
- RT328T
- RT528A
- RT528E
- RT508A
- RT517
- FR540
- RT514R
- RT515A-1
- RT385
- RT485

The following Bendix/King radios can be replaced with TKM Michel slide-in radios:

- KX170 series
- KX175 series

Narco Avionics can provide replacement radios for the following:

- COMM 11
- COMM 11B
- COMM 10
- COMM 111B
- COMM 11A
- COMM 110
- COMM 120
- COMM 111

The new radios comply with the new 0.003-percent frequency-stability requirements, and most don't require any modifications for installation, but you'll need to check to be sure. The only remaining problem is that the wiring is still the same and a potential hot bed for failure at some point. Any decision must be tempered with the knowledge that failure may still occur, maybe not in the radio, but in the wiring.

TKM Updates and Modifications

The following are updates and modifications that can be applied to your TKM radio should you experience one or more of the indicated problems.

FM'ing in all TKM radios

Resolved with introduction of new communications synthesizer.

MX 170B.

170B Audio Feedback

Refer to Installation Bulletin No. 56 from King Radio for the KMA-20 and KA-25. The following is extracted from Bulletin 56 for your convenience.

> *Subject:* Audio Bleed Through When Using KMA 20 or KA 25 (date: 06/28/74)
>
> Field experience indicates that when two KX 170 series units are used with a KA-25 or KMA 20, audio from the "unused" communications receiver can bleed through and be heard along with the "in-use" Communication receiver audio. This audio level is considerably weaker than the selected channel and is not usually noticeable with engines on. The reason for this occurrence is the audio power, which is developed by the KX-170 series communications speaker amplifier and is dissipated, in a loading resistor located in the KMA-20 or KA-25. These loading resistors are located in the isolation amplifier as a convenience. However, if the KX-170 series communications receiver volume control is set above 50 percent, 5 watts or more, audio power is being dissipated into the load resistor in the isolation amplifier. This 5 watts of audio is enough power to "pump" the ground on the loading resistors causing the undesired audio to be heard in the selected audio channel. (A good clean bond to ground is necessary for the loading resistor.)
>
> To eliminate this condition, it is recommended that the speaker lines from the KX-170 series unit be routed separately to a load resistor mounted external to the KMA-20 or KA-25. Keeping the communications volume control at a lower setting will also reduce the audio bleed through.
>
> KMA-20, S/N 11233 and above have the loading resistors relocated internally so as not to exhibit the bleed through symptom. However, this modification requires mechanical changes so there will be no field or factory modification available. This situation is considered a nuisance factor and not a serious operational problem.

Multiplexing

When multiplexing occurs, intensity variations cause a trace of high-pitch tone that may be heard in the audio. Modifications have reduced the noise in the audio circuit.

MX-385 Power Failure

This is caused by excessive random voltage spikes. Resettable fuse modification installed into the power supply circuit prevents this condition.

Receiver Performance

Development of new navigation and communications receivers have eliminated unreliable inductors and reduced FM interference.

MX-170B Mounting in Tray

This may cause weak receive and transmission. It is corrected with an extension of the rear panel.

TKM Radio Shaft Breakage

This occurs when die-cast shafts are replaced with stainless steel. Should technicians or end users encounter problems with TKM radios, they may contact their respective dealers or distributors. Such problems should be written and faxed or mailed for review. The intention is to improve the quality and reliability of TKM products.

Besides maintenance and reliability problems, one of the reasons you as the customer/pilot might consider replacing old equipment is simply all the new bells and whistles that come with new avionics. The important reason is safety in the form of reliability and accuracy. If much of your business depends on the availability of your airplane, then you'll be sensitive to the need to have radios that are subject to minimum downtime and provide accurate, reliable service.

Used Radios

Could used radios be a good choice in lieu of new? If so, your best bet might be to buy an entire package that was removed from another airplane, not a mess of miscellaneous boxes accumulated from several different airplanes. A system that functioned well as an integrated package has a greater potential of working reliably after being installed than a hodgepodge system and, furthermore, shouldn't be a warranty nightmare. The worst thing you can do is attempt to put together an avionics system consisting of various radios from different manufacturers. If you have to buy used radios, make sure they are all of one make, from one manufacturer. The reliability and savings in installation time will be more than worth any extra cost.

There are various precautions to watch for when considering buying used radios. Perhaps the previous owner had an intermittent problem with them and wanted to get rid of them. Have all the factory modifications been done to keep the radios up to date? Will you be allowed to return the radios if they don't pass testing at your avionics shop? Does the used radio have all the features you need?

With used avionics more than ten years old consider that many parts might be difficult to obtain. Even if available, they probably have significantly high cost because of low availability. Will the pilot have the time to wait for replacement parts to be shipped from an obscure source that hung on to its entire old inventory?

What do you as the customer need to know to keep from making a major financial mistake? You need to be familiar with how the stock, repair, and return-to-service system works. Tagging parts is extremely important because all you need to know about the part is most likely available if you know what to look for. The so-called "yellow tag" is a method used by repair stations to identify items that have been repaired and officially returned to service. The color yellow, despite what you may have heard, does not connote any special kind of status to the repaired or overhauled part. Yellow had simply been settled on as a convenient convention; the tag could be any color at all. An important consideration when buying used, yellow-tagged radios is that the FAA now requires that all documentation pertaining to a particular repair must accompany the repaired item. Most repair stations use yellow tags, and if you read the tag you'll see a note that references a work-order number. For a repaired or overhauled radio to be legal, the actual work order must be attached that lists the parts used, what the technician did to the unit, and how it was tested after the repair. You'll find that this is rarely done, and you might have to insist, but do so. It's important that you know the pedigree of your avionics, and this knowledge will come in handy if you ever sell your airplane. If the yellow tag doesn't have a work order attached to it, don't buy the unit, no matter how good the price.

Detecting Stolen Radios

Beware of stolen radios. Expensive products always bring the unscrupulous out of the woodwork. One way to check for stolen items is to check serial numbers against the stolen radio listing compiled by the Aviation Crime Prevention Institute (ACPI). Most radio shops subscribe to this list, so they should be able to provide one to you.

Radio thieves have a few tricks up their sleeves, however, and you need to be watchful of their skullduggery, as both an owner and potential avionics buyer. Their cleverest trick nets them a set of stolen radios that no one ever suspects is stolen. Here's how it works: They fly into a dark, unattended airport in the middle of the night. They have previously checked out the airport and pinpointed the airplanes they are going to rob. The robbers open the door on the first airplane; let's say it has a full panel of King digital-display avionics. Opening the door is simple, by the way. Airplane locks are incredibly cheap, and a set of about fifty keys will open most door locks on most airplanes at any airport. They remove all the panel-mounted avionics using a hex-key wrench. This takes about three minutes. They leave the door unlocked.

At airplane number two, which has the exact same radios, the robbers unlock the door and remove all of its radios. Then they install the equipment taken from airplane number one into the panel of airplane number two, and lock its door after carefully removing any traces of their illegal entry.

The robbers take off. The total time for the switch is about ten to fifteen minutes maximum. Next morning, the owner of the airplane number one gets a big surprise: All the radios are gone. Naturally, the owner files a report with the police and insurance company. Being conscientious, the owner kept a record of the stolen radios' serial numbers.

The owner of airplane number two never notices (s)he is flying with a different set of radios, unless at some later date (s)he has reason to note serial numbers and compare then with those of the original radios. Perhaps (s)he will think the avionics shop mixed them up.

Eventually the serial numbers for airplane number one's stolen radios gets on the ACPI list. Yet, those radios will probably never be found because they are sitting in airplane number two's instrument panel. In addition, someday an airplane owner trying to save some money will end up owning the radios that were stolen out of airplane number two's panel because those serial numbers are still "legit."

Are stolen radios that big of a problem? The ACPI reports that 158 pieces of avionics were stolen in 1992. This figure has dropped for the fifth straight year, so fewer radios are being stolen. However, if you can afford to, buy new radios, rather than used, to avoid the possibility of purchasing stolen goods.

Antitheft Strategies

Airplanes are so easy to break into it's pitiful. Moreover, many airports can't afford night security, so what are you supposed to do to protect that $15,000 instrument panel? The best solution (and the most expensive) is to keep your airplane hangared. This isn't always possible, especially when away from home base, but there are other options.

Because airplane locks are so easy to open, the first line of defense is to install a better lock. Barrel-type tubular key locks are available that fit most light airplanes. The easiest way to replace the lock and find one that fits is to remove the old lock and take it to a locksmith so you can be sure of getting the correct size. Remember that you'll have to do this under your mechanic's supervision to make it legal. A barrel lock on a paper Cherokee cabin door and baggage compartment will make the airplane very difficult to break into unless the robber chooses to break the windshield. Curiously, many airplane robberies result in little or no airframe damage.

You can install barrel locks on Cessna singles, but why bother? The side windows on Cessna singles are easy to open, so having a door lock is somewhat silly. In many cases, all you need to do to open a locked Cessna door is to simply rock the wings: The pilot's door, if not adjusted for a very tight fit, will simply pop right open.

First off, keep all windows covered with sunshields at all times when not flying. This will keep people from admiring your instrument panels and all the pretty toys. Besides, this will also help protect it from the unrelenting effects of the sun.

Second, consider buying one of those thick, metal instrument-panel covers that lock in place when you're done flying. You'd need some special tools to get one of these off, and most robbers want to get in and out quickly so they won't take the time to deal with a good panel cover. The other benefit of these covers is they prevent people from seeing what's in the panel in the first place. Don't install a car alarm. It would be a paperwork nightmare to start with, plus any kind of wind can set it off. If you insist on having an alarm, get one designed for aircraft. There are some pretty nifty systems that will destabilize your sense of balance because the horns are so loud. If you have the money, you can get movement sensors that detect the plane's motion and even cameras to record the dirty deed.

Here's another scenario that will make your wallet lighter. You signed the bill of sale, the airplane is yours, and you fly it home. Everything checked out okay on your test flight except for a few minor items that the owner agreed to fix before you picked up the airplane. Surprise! On the way home, the avionics system fails. What happened?

The seller pulled a switch, replacing the radios that were in the airplane during the test flight with defective or older models. It's hard to believe, but I've seen it happen, and it probably will happen again. In the particular case I'm familiar with, the buyer hired an attorney and was able to recover damages, but how about you? Will you check close enough before you pay for the airplane? Have a dependable shop check out the plane before purchase.

New Radios

Rather than installing "experienced" (used) equipment, check out the cost of a new avionics package. You'll enjoy a new warranty, sometimes as long as two to three years depending on the manufacturer, plus all the features you'd expect from modern avionics.

Regardless of the equipment selected, your choice should be guided by determining what will provide long-term reliability. This includes not only the avionics itself, but also the shop that installs it. In other words, does the shop keep stock and provide loaners for the types of radios you're thinking of buying? Also, how long will the shop provide warranty for its installation, and what is covered by its installation warranty? You may pay less for a particular radio, but to what purpose if its subcomponents are subject to frequent failure or if the shop sells it so cheaply they can't afford to support you after the sale?

It's important to find out the terms for warranty repairs before spending thousands of dollars on new radios. You might be interested in knowing, for instance, that Narco Avionics no longer sells parts for current production radios (and some older radios that use the same parts) to its dealers. Narco wants the dealer who sold you the radio to send it back to Narco for repair. Narco says its turnaround time is about three days.

Another major benefit of replacing old avionics with modern equipment is that the new installation can free up a lot of instrument panel space. You might find that, in considering a new package, you could add a piece of equipment like a GPS where before there wasn't room. Another option that provides that extra bit of insurance is the VNS-1000; a device that simulates a GCA (ground collision avoidance) landing system approach. In less than four hours, the VNS-1000 unit is installed and operating. Using human speech, the VNS-1000 provides commands that instantly keep the pilot informed of deviation errors. Commands such as "System Armed, Front Course Selected, Back Course Selected" and correction commands in degrees are immediately verbalized for the pilot. If the plane is off by one degree right of the runway, the VNS will state, "Correct one degree left." The commands will continue until the plane's course is corrected.

The "glass" indicators available today for general aviation can upgrade the avionics in your plane, providing that little extra measure of reliability and ease of use without having to change out all the avionics. For example, the Avidyne mulitfunction display replaces some of the older radars with a glass, electronic display that can present several different indicators, including the radar.

There are many advantages to installing new equipment, but there are still potential problems. Planning and good communication with the avionics shop manager during the negotiation stages will reduce the obstacles you are bound to encounter.

I like a good deal as well as the next person, but I'm not enamored about the mail-order business for avionics. I think that ordering by mail-order is a risky business when dealing in aviation radios. If you're lucky and everything works, you're probably in good shape. However, when you consider the potential problems that normally come as part of any avionics package, do you want to take on the additional hassle of having to deal with a long-distance supplier?

Other Purchasing Considerations

Deciding what brand to buy is a personal decision. You may be guided by advice from other owners, the fixed base operator (FBO), or by doing your own research. Before you get close to deciding, however, you might want to start reading some of the avionics magazines listed at the end of this book. In addition, *The Aviation Consumer* magazine has conducted extensive reader surveys of avionics reliability. You might try to read some of their back issues to help in making your decision.

If you are an avionics technician, you enjoy a unique relationship with the pilots; you work very close with them on their squawks and know what avionics system has demonstrated the most problems with the highest costs. Help the pilot to better understand what to look for in deciding on a new avionics package or adding a modification. You can do this by providing the pilot with brochures, excerpts from Advisory Circulars, and the like.

Technical Standard Order (TSO) Requirements

Avionics that have met TSO requirements, which are a set of minimum standards, may add more resale value to your airplane, especially if it is a light twin that might be used someday for charter. In addition, the FAA is pushing for all new installations to be approved radios (read TSO'd, STC'd, etc.).

Cooling Fans

Avionics cooling fans are essential with today's jam-packed electronics. Even though modern radios don't contain tubes anymore, to get the fantastic capabilities out of these new radios, circuit boards are densely packed with heat-generating components. For proper operations and long life, it's crucial to get cooling air flowing through the radio stacks to dissipate heat.

Ambient-air temperatures can adversely affect avionics equipment, severely limiting service life, even when not turned on. Congestion behind the panel, inadequate venting through the glareshield, and the greenhouse effect of an unprotected airplane sitting outside without windshield cover all contribute to reduced avionics life.

Any temperature buildup beyond the recommended operating range (the maximum is 131°F for TSO'd radios) will begin the destructive cycle. For each 18-degree rise in temperature, the life of solid-state circuitry is reduced by 50 percent. Although the radio might continue to operate, failure could occur without warning during a later heat wave or become intermittent because of a component that weakened from excess heat.

In older radios, recurring failures can be caused by heat-sensitive components. Wax-filled capacitors used in some old radios, for instance, spill out their innards when subjected to intensive heat behind the instrument panel. Normal operation levels deteriorate as components grow old, their insulation dries out and cracks, and higher voltages sneak out through these escape routes. One repair might get the radio going again, but then a domino effect takes place. The new, stronger component is installed, and the weaker, older components in the same circuit absorb the new load allowed by the new component and finally another part dies. This process can continue for months, and at some point a smart avionics technician will simply suggest replacing one of these radios with a new one, rather than replace all the components one at a time.

Avionics Cooling Fan by Troll Avionics, Inc. Model FN-200
Unit is FAA PMA'd, filtered and shielded

Avionics cooling fan. Up to three hoses can directly connected without a "Y" with this fan. Fans are available with up to six outlets, but airflow on each may not be the same.

Some installers connect a cooling air hose from a scoop on the outside of the airplane to the radio stack. While that is better than nothing, a better solution is to install an avionics cooling fan that runs any time the master switch is on.

If you are having a new radio installed, make sure the radio stack has room for a cooling plenum. A plenum is a plastic chamber that attaches to the radio rack and directs cooling air onto each radio in the rack. Hoses from the cooling fan can be attached to the plenum.

The cost of a cooling fan is much less than the repair bill for heat-damaged radios and is worth it in terms of the reliability of the equipment. To provide adequate airflow, at least 0.002-inch spacing should be left between radios stacked on top of each other.

Reinforced 5/8 inch Avionics cooling fan hose.

Cooling fan connecting hose.

There is no question that reducing the operating temperature of electronics components dramatically improves their life expectancy. General aviation avionics systems have to repeatedly endure one of the harshest environments of any electronics package: the avionics panel-mounted stack. Not only are they crammed tightly against other radios in the same stack, but they also suffer from what is called the chimney effect. This is where heat from the bottom radios rises, gathering heat from each radio and spreading throughout the entire radio stack, increasing ambient temperatures far beyond designed specifications.

With the addition of small, but highly effective direct current fans that force cooler air across the entire radio stack, this cumulative effect is reduced and even eliminated. To make matters worse, from a position of potential heat damage, design engineers are using more and more surface-mounted devices (SMDs) to reduce size and weight. The fairly recent Sandel 3-inch display and the even more recent McCoy IND-2000 navigation indicator (NAV/G/S/GPS) are good examples. Although these devices use less power, they are more sensitive to heat. Adding a cooling fan is perhaps the best decision even the cost-conscious aircraft owner should make. The installing facility should refuse to install a radio unless adequate cooling is provided. Cooling fans stabilize the ambient temperatures behind the instrument panel and keep the operating temperatures well below the potentially damaging levels that can destroy radios. Some levels of damage may not be so obvious; the level of degradation may not be immediately seen as a failure. As vibration and continuous changes in temperatures persevere, more and more failures may occur and be difficult to find. These will most likely be written off as NFF (no failure found).

How much does anyone who installs cooling fans really know about the part he or she is installing? Not much, but the difference can make or break the life of an avionics radio. Everything from bearings, filtering, human error, and design can affect reliability of the fan. Fan failure may not be detected and the heat within the avionics will escalate, eventually causing damage or error.

Brushless DC Fans

Brushless DC fans for aircraft are available with three voltage capabilities: 12V, 24V, and 28V. Because the speed and airflow of a typical DC fan is proportional to the voltage supplied, a single product line can be used to meet different heat-sensitive applications by providing a supply voltage that will give the desired airflow. In other words, the higher the voltage, the faster the fan turns and the more air it moves. Lower voltages move less air, so the decision to pick a fan is determined by the amount of air needed to be move.

The voltage range necessary to provide satisfactory operation depends upon the fan design and may be anywhere from 10 to 14V for 12V units and anywhere from 12 to 28V for 24V units. Brushless DC fans do not draw constant currents. Throughout the operational cycle, the currents being drawn will vary from their minimum to maximum limits. Since DC fans have much higher starting torque that their AC counterparts, the time to reach full speed with the use of current limit will be less than equivalent AC models.

Electromagnetic Interference (EMI)

Engineers are striving to meet the EM emission requirements set forth by the FCC. Brushless DC fans are not excluded; the potential for various forms of EMI are real and must be considered, both by the fan manufacturer and the avionics engineer and technicians.

Attempting to isolate and define this aberrant interference can be difficult, especially when considering the varying perspectives taken by those that design and those that have to troubleshoot.

All of those that work with aircraft avionics use the term electromagnetic interference (EMI, also known as RFI [radio frequency interference]. This describes the undesired interference energy conducted as currents or radiated as electromagnetic fields. The EMI current within brushless DC fan-power leads are described as "conducted" EMI (450 kHz to 30 mHz) and is considered to be more of a problem than "radiated" EMI.

Electromagnetic Compatibility

Electromagnetic compatibility (EMC) is best described as the capability to operate without generating unwanted electromagnetic interference that can interfere with the operation of other equipment, and does not respond to undesired externally generated interference. If these problems are addressed during the design process, it will save expensive solutions later.

Magnetic Fields from DC Fans

Most brushless DC motors have the magnetic fields of the permanent magnet and the stator windings encased within steel housing, which provides a degree of shielding of the magnetic field. For the most part, mounting the open side (inlet/visible blade) fan away from sensitive electronic circuits and CRTs like that in a radar display unit should provide a measure of protection from the possible magnetic fields that may be produced by the fan motor. A quality fan should not inject any problems into the avionics.

The primary mode of fan failure is bearing damage caused by the lubricant degrading over a period of time. Elevated levels of temperatures that can occur behind the radio panel can accelerate the failure rate.

Selecting an Avionics Fan

The choice should be based on initial quality requirements. If you desire a quiet fan from the onset, use a fan with a sleeve-bearing configuration. Initially quieter, but over the long haul with increases in ambient temperature under the glareshield, it will demonstrate much higher noise levels than its counterpart, the ball-bearing fan.

The ball-bearing fan will be noisier in the beginning; the pilot will hear an obvious hum as it starts up and runs. However, as temperatures increase, the bearing design will not appreciably increase its noise level.

Comparing apples to oranges, generally speaking, there is not much of a life differential between a sleeve-bearing system and its equivalent ball-bearing system under total low temperatures, but as this total temperature increases, ball bearings give progressively longer life than sleeve bearings. Handling of the bearing fan must be much more gentle; they are sensitive to shocks that affect the individual points where the balls make contact. Sleeve bearings spread that shock over a larger surface area.

Conclusion

So, what have we learned about selecting a fan? Although ball-bearing fans appear to have longer life, they most likely will be noisier and cause far more noise complaints than their sleeve-bearing cousins. The ball-bearing versions last longer, but as they begin to wear, the shrill noise of irregular bearing surfaces rubbing at high speeds becomes unbearable, to excuse the pun. With this in mind, it is not surprising that selecting the sleeve-bearing fan may be the best choice. Also, make sure the fan is PMA or TSO'd and the certification is actually for the fan and not for some forgotten installation.

Installation

A lot of discussion revolves about the technical aspects of fans, but there still remains the physical installation and some expansion as to the need for the fan. There are several ways FBO (Fixed Base Operator) shops will install cooling fans in and around the radio stack. One method where air vents are not provided is to mount the fan directly on the bulkhead and vent the flow of air with hoses directly to the back of the avionics stack. This would require fasteners to be drilled through the bulkhead and sealed. Drilling through the bulkhead is not to be taken lightly, especially on a pressurized aircraft. Although this will provide some relief, a more direct and more efficient method is to create a plenum that will mount directly to the side of the radio stack. Only one input hose is required to force air into the plenum. The side of the plenum that faces the stack is designed with 25 or 30 holes that allow the flow of air to be forced between and around the radios. Newer radios are designed with an input fitting that accommodates a single hose from the fan. Each radio that has a fitting will have a standard 5/8-inch CAT type hose connected. A "Y" splitter allows a three-port fan to accommodate six radios, but it requires more work than a six-port fan. There will probably be cases where the fan will be mounted directly to the radio rack. This should not be the norm; there are other radios that need attention.

Three- and six-port units fan designs are the most common in general aviation. When selecting a fan, look for one that will provide the same output from each port regardless of which is plugged or used. Another feature to look for is polarity reversal protection and physical size. Finally, is it truly TSO'd or just using a general-purpose tag to obtain validity for the purpose of marketing and sales? Actually, it would be best to have a PMA certification. Make sure the CFM (cubic feet per minute) is 26.6 or above and that physical mounting will not warp the case to the point of hindering blade rotation. Is there only one mounting plane? The mounting points should allow vertical or horizontal positioning with no change in fan speed.

The cooling fan should be mounted to take ambient cabin air and force it in and around or onto the radio stack. If the radios are equipped with air ports, installing and applying cooling air with a fan is fairly easy; it doesn't take a lot of volume to cool the radios, just a steady flow. In no case should an outside scoop or forced air be used; moisture and air laden with dust can severely damage the radios. Besides, scoop air isn't available when the plane isn't moving.

The fan can mount somewhere under the panel, behind or on the bulkhead. It doesn't make much difference where, as long as the fan has an unobstructed area directly in front of the open side (read inlet) of the fan. There is some disagreement on where and when the fan should get its DC power. One school of thought is that the fan should be wired directly to the avionics buss. This way when the avionics master is turned on, the fan will run and start cooling by moving air away from the stack. However, another group wants to see the power connected directly to the main aircraft buss. The rationale for this is that by the time you turn on the avionics master switch, the voltage on the buss is stable, theoretically allowing the fan to last longer. It is more likely that this won't make much difference one way or another. The fan has a wide operating voltage and isn't that sensitive to spurious voltages. Actually, they are supposed to be designed to handle the real-world voltage conditions found on the avionics buss.

The avionics cooling fan motor should be a RFI/EMI tested motor, filtered and shielded to prevent interference with the aircraft radios. It is simply not worth installing an inexpensive fan and having to fix the resultant problems.

Look for fans that have the approximate specification as follows:

- Operating Voltage: 14 to 28 VDC (depending on aircraft power)
- Current Draw: 450mA to 490mA—startup will be almost double
- Size: 4.88" × 4.88" × 1.4" or smaller
- Weight: 17.4 oz. or lighter
- Certification: PMA, DO160C, or at the very least, FAA TSO
- Complete self-contained with color-coded power and ground lead
- Shielded and filtered
- Mounts directly to tray or other available structure
- 11 to 32 Volt operation
- Two or more years warranty
- Price between $150 and $250, depending on degree of reliability and warranty period.

CHAPTER 4

Avionics Installation and Repair

Developing an Installation Plan

When a new installation or modification is planned for a customer's plane, the pilot and FBO Avionics Manager should develop the installation together. Once they have arrived at an acceptable plan, the installers and technicians will be presented with the proposed installation. They will input their concerns and expectations, which will be incorporated into the total estimate. At this point, the customer will be appraised of the predicted cost. Once the package is approved by the customer, the plane is then scheduled for installation.

While the pilot won't be directly involved in the installation unless building his or her own airplane, he or she should discuss the following points with the avionics shop manager to be sure they are taken into consideration during your installation. These factors will go a long way towards increasing the reliability and maintainability of your installation. If you are the pilot or owner and building your own airplane, use these points as guidelines for your avionics installation.

One of the most important factors to consider during an installation is designing the installation for maintainability. When access is easy, technicians are more likely to perform quality repairs. But when technicians must battle for hours just to reach the problem area, their decision-making ability could be strained and dampened with sweat. Besides the effort in accessing the avionics, damage to surrounding wiring and connectors could occur without the technician's knowledge. Fixing one problem and creating another doesn't help you, the owner. From the outset, then, designing for maintainability is a worthwhile goal.

The second goal, and one that is also important, is design for reliability. This involves working to high-quality standards to prevent problems caused by vibration, moisture, chafing, RF interference, and other sources of trouble.

Installing the Avionics Radios

The initial part of the installation will involve building wiring harnesses, then roughly routing the harness into the airplane to see how it will fit. During the rough routing, the installer should note all points that will require special attention to avoid chafing and RF interference. As the actual routing takes place, the installer should take care of those potential trouble spots by installing clamps, protecting the harness with plastic spiral wrap, and installing caterpillar grommets in lightening holes. To avoid RF interference, the harness should be routed to clear high current cables and any other wiring that could interfere with each other electronically.

To prevent chafing, all wiring harnesses, plumbing, and installed equipment should not come into hard contact with the structure or with each other. At no time should equipment or equipment racks come into direct contact with nearby structures. Because the aircraft skin and structure can flex during flight, establish at least a quarter inch or more for adequate clearance around the proposed and adjoining equipment.

If the wiring harness is lying gently on smooth aluminum skin with no sharp edges, the likelihood of damaged wiring is slim, but possible. The key is relative motion and pressure. The greater the pressure and relative motion, the greater the potential for insulation breakdown. To be safe, it's preferable to clamp the harness to protect it from rubbing against the skin.

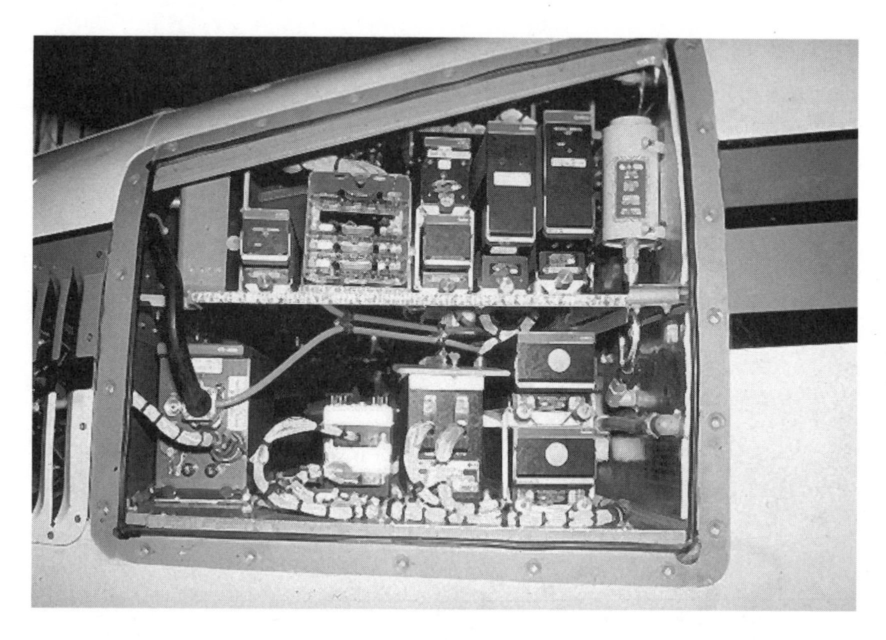

Avionics bay (nose of aircraft). Note the well tied-out harness with strategically located clamping.

Shock-mounted equipment racks must take into account the movement of the radio during normal aircraft operations. The same is true of wiring harness clearance from plumbing lines, especially those carrying oil, fuel, and oxygen; there should be no physical contact between adjoining plumbing, wiring, or structure. Never use plumbing for primary support. It's okay to use standoffs to separate harnesses from plumbing, but at no time should the plumbing carry the weight of the harness.

Wires are insulated, but the insulation isn't impervious to damage from sharp edges, heat, or excessive pressure such as that imposed when wires are clamped with nylon ties against a metal surface. For more detail on important harness-routing considerations to avoid RF interference, see the tips section within the respective chapter for that particular piece of equipment.

Designers must work to a consistent standard for harness installations that takes into account all these factors, plus what I call the "five don'ts."

1. Don't limit clamping provisions. A sufficient quantity of clamps is necessary to prevent harness droop between clamps.
2. Don't route harnesses to come into contact with sharp surfaces or ride against any moveable surface. Provide antichafing if necessary.
3. Do not design location and space requirements without allowing for service loops (adequate slack in harness that will allow maintenance, such as changing the connector up to three times).
4. Don't design harnesses to route in hot areas without adequate thermal protection.
5. Don't route harnesses in areas that are subject to chemical damage without protective conduit, such as landing gear wells and engine compartments.

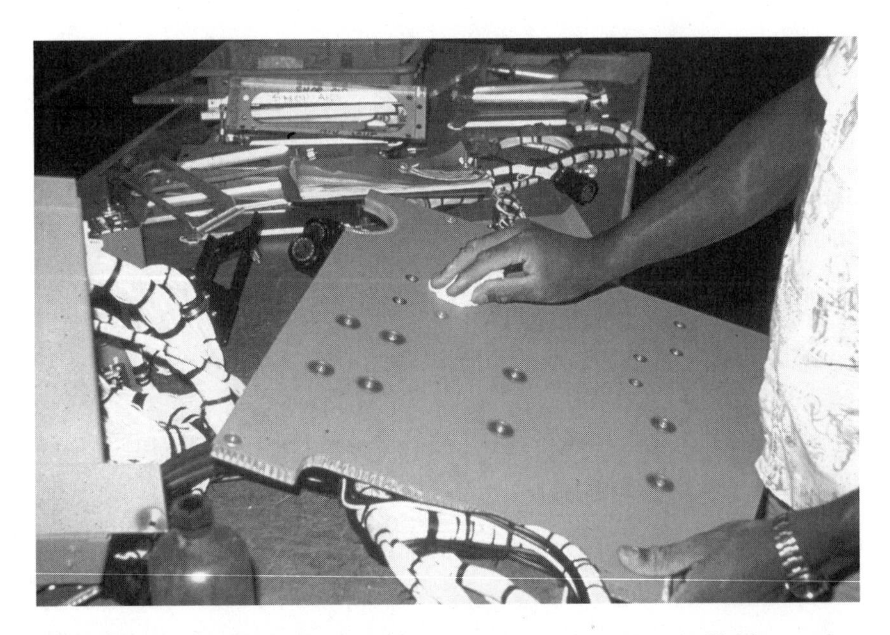

Rack mounting. **The shelf is prepared with through-shelf inserts that are first drilled, edge-filled for strength, inserted, and injected to prevent the inserts from coming loose and to add additional strength.**

Follow these rules on mounting radio receivers, transmitters, amplifiers, and computers.

1. Hard-mounted equipment can be installed as close as necessary to other equipment except for clearance needed for cooling (usually 1/4-inch). Rack-to-equipment contact should take bonding into consideration when depending on tension contact. Paint should be removed from the radio where tension contact is expected to touch bare metal.
2. Provide harness supports at the back of the rack to alleviate stress on wiring and connectors that could cause difficult-to-troubleshoot failures down the line.
3. Allow sufficient distance between the radio and the aircraft's skin. Normal airflow and aerodynamic stresses on the skin can

Note harness seals where the harness passes though non-pressurized bulkhead to next section.

cause changes in this clearance. The goal is to avoid contact between the skin and the radio.

4. Fasteners for holding radio racks in place should be secured with locking devices (either a lockwasher or locknut) to prevent vibration from allowing screws to loosen. This is especially important where radio racks sit above flight controls. If a loose radio rack could impinge on flight controls, it is a good idea to add supports to the rack as a backup to prevent flight control interference.

5. Provide sufficient space for wiring harnesses and coaxial cable connectors. Coaxing cables must enter the mating connector on the equipment in as straight and natural a routing as possible to prevent connector damage.

6. Provide protection from moisture. The time to find out your windshield is leaking is before you spent $10,000 on a new radio installation. If you suspect a leak, spend some time under the instrument panel with a flashlight while a friend sprays the airplane with water.

7. Make sure all fasteners used in the installation can handle stresses that will be imposed.

8. If the installation is in a pressurized aircraft, all areas that penetrate the pressure vessel must be sealed to prevent cabin pressure leakage. One small leak might not affect pressurization, but a number of small leaks could cause a significant drop in the ability to pressurize the cabin.

Instrument Panel Manufacturing

Designing an instrument panel or adding a new radio demands a great degree of individual responsibility. Engineers aren't the only ones to lay out and install a new avionics component or system. In the field, technicians usually have that burden. Pilots rely heavily on the foresight and planning that goes into designing the instrument and radio panel. In today's aircraft, more than ever, switches and indicators need to be readily accessible and easy to find. There is an understandable desire, and yes, necessity for engineers to allow for future additions of equipment. The problems arise when that new switch is added and placed where it is neither ergonomic nor functional.

Emergency situations have led to many pilot errors, like throwing a switch on some Beechcraft models to select cowl flaps and instead activating the landing gear. Their location can be confusing to inexperienced pilots. You can't very well land smoothly with only the cowl flaps extended; they make for poor landing gear. Pilots simply forgetting to put the gear down after several complex go-rounds is yet another problem, but this type of issue is a lot harder to fix. Only experience will help these pilots, or the addition of an aural warning device that immediately alerts the pilot if the gear is not down and

Installed instrument panel.

locked during landing. Factors that can be controlled are the proper and clear marking of safety of flight controls and switches that are ergonomically located. Keeping the plane centered on the runway is yet another problem, but this is easily resolved with a device that verbally instructs or warns the pilot to correct to the proper number localizer needle degrees to bring the plane back to center of the approach. Landing is a very stressful part of flying for pilots. The FBO and their crews can make it less strenuous by maintaining the plane in top notch shape.

There are two objectives for adding instruments, switches, and monitor light assemblies to an instrument or avionics radio panel. One is the location and the other is how the installer will accomplish the task. Both are equally important, but the one that will cause the most uproar is a sloppy job. The pilot has to look at the panels and mistakes left by the installers during the entire flight. So, *don't* make them! The following paragraphs cover some of the concerns from pilots relating to equipment location, plus installation tips. The suggestions covered don't take into account all the new tooling systems now available. Not all shops have the finances for such toys. Armed with only a minimum number of tools, the qualified installer can create a professional installation that easily equals factory standards!

This doesn't mean that choosing a shop to subcontract the panel work isn't a good idea. In fact, it may very well be the best avenue open to the modification center. For example, a well-equipped panel manufacturing and modification shop is more likely to be experienced in creating a long-lasting panel that will wow anyone who sees it. Besides, any errors are the final responsibility of the shop. Werner Berry Aircraft Engravers, located in Ontario, California, will create a 1/16-inch thick plastic overlay that will cover your original panel or create a brand-new one. The advantage of this approach is none of the placarding can be rubbed off or damaged. It is reverse engraved and can be brightly colored to highlight the respective lettering, like call letters.

Another advantage is the mounting screw heads can be completely hidden under the overlay, if desired, revealing only the indicator

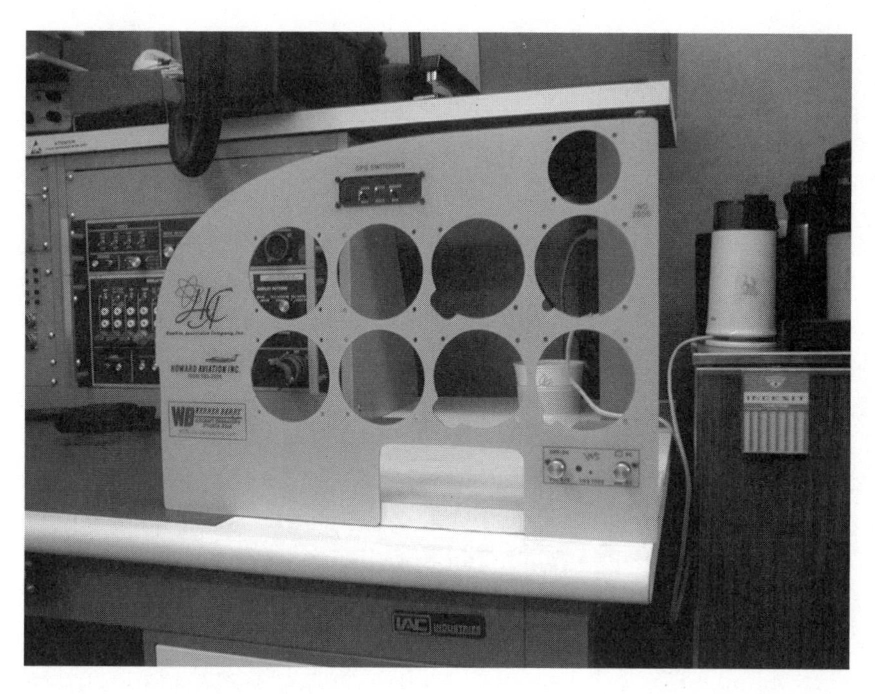

Pilot panel with 1/16-inch overlay by Werner Berry.

face. This allows for an improved appearance for the turn and bank (T&B) indicator where the holes may be slotted. When this overlay is viewed for the first time, it is virtually impossible to see that it is an overlay. Fingerprints no longer mar the paint. The final results are a photo finish for the client and the shop. Two other shops provide panel restoration or overlays. Paramount Panels, Inc. manufactures and refurbishes aircraft panels of all types, such as circuit breakers and switch panels. They are also located in Ontario, California. The other company is E.D.N. Aviation, Inc. They are an authorized FAA repair station. Their specialty is panel engraving and manufacturing. Another option that is open for some applications is heavy-duty, self-adhesive placards that can easily be created using any art work processor and an inexpensive light-developing process. Vital Presentation Concepts, Inc. (VPC) can provide not only the placards but also the equipment for a shop to do their own. The process is easy and relatively inexpensive.

Refurbished edgelite panels by Paramount Panels, Inc.

With any of these companies, it is recommended to have a signed agreement to correct any errors they incur without additional charge. This would also be helpful to assure yourself that everything is as your customer desires and matches the designs from engineering. Rework can be expensive, except for the VPC product, which is not only inexpensive to manufacture, but easy to apply.

Should it be decided to do the work in-house, a lot of soul-searching is suggested. Making mistakes can be very expensive. Be prepared to redo the panel several times if the client should change their minds. Don't get frustrated, just make sure the client will pay for the changes. Know what you are doing; practice with scrap metal and plastic before attempting to put real money and time into making an instrument or radio panel.

Getting past the why and when is where and how. Burrowing holes in the instrument panel or the radio panel without a lot of experience takes either a lot planning or being just a little crazy. Make one slip and a perfectly good, well-designed, painted panel becomes your worst nightmare. Not only is the installer subject to the anger of the supervisor, but also the plane's owner or pilot. Shop owners aren't in the business of losing money. The bottom line is not to consider cutting indicator holes or cutouts for radios unless it is a regular part of your job. Work with the shop's sheet-metal person, and continue to improve your skills until they are second nature. To repeat, practice on scrap metal before cutting into the customer's panel. Even then, it requires planning to anticipate potential disasters such as interference with plumbing, structure, or adjoining equipment. Templates and specialized tools are available to assist the installer in making accurate cutouts for instruments and light assemblies. These are available from various distributors. PC-based software programs such as Panel Planner are available for a reasonable cost and allow the shop to lay out the proposed panel and present it to the client. Additionally, it opens the door to other possibilities such as previously unseen problems that could not possibly be considered without a picture to compare with.

Weather conditions, long trips, exhaustion, and being anxious to get home can soon play a major part in judgment calls such as failing to select the VOR to the local field, or confusion with ATC directions. Making the right decision can be further aggravated when there are seemingly rows and rows of avionics and electrical switches dotting the instrument panel. Key switches such as the communication and navigation audio switches could be coded to accentuate their importance and priority. What is more important is not to locate them close to safety of flight controls. I cannot state this strongly enough: The technician and installer must be aware of the problems affecting the pilot *before* arbitrarily punching holes in the panels.

The pilot's attention during any given flight is divided between navigation, communications, aircraft systems functions, and his or her performance. Much of what the pilot does is semiautomatic as a result of training. The less experienced the pilot, the more he or she must work to keep abreast of what is happening at any given time. Throw into the mix all the diversions that occur during many flights such as a change in weather, equipment failure, or new instructions by ATC, and therein lies the ingredients for disaster.

The conditions that affect the single-engine pilot are similar to those that fighter pilots encounter during combat conditions. An overwhelming mass of data being thrown at the pilot at one time will essentially causes sensory overload and mistakes begin to happen. These mistakes can be fatal! What fighter pilots will do when this condition occurs is shut down everything but that which is absolutely necessary to complete the mission.

Biotechnology studies the pilot's environment, and attempts to provide the best possible scenario that will make for a safe flight. Factors that continue to affect development of a fail-safe cockpit are the many obsolete avionics systems, replacement radios, added equipment, and wiring changes. These factors force designers and avionics technicians to place switches in less-than-desirable locations. In the real world, this is what most likely happens, but it would be great if a new panel could be created to provide the optimum layout.

Older planes with only a single communication radio and one navigation receiver are an example of this approach to adding equipment. Audio switches should follow the conventional configuration from left to right: COMM 1, COMM 2, NAV 1, NAV 2, ADF 1, ADF 2, DME 1, DME 2, MARKER 1, MARKER 2, etc. What happens is other radios are added, plus the respective circuit breakers and switches. These are usually placed out of sequence rather than being properly relocated. This is simply a matter of cost and must be avoided. If the cockpit lights fail, the pilot should be able to find switches by feel alone without a flashlight. For this reason, under no circumstances should the communications or navigation switches or breakers be changed from the recommended sequential layout.

The standard "T" layout of instrumentation is fairly common and is still recommended by mainstream professionals for pilots to scan and understand what they are viewing. The only catch to this is some less experienced pilots may focus only on the flight director or navigation indicator on approach and ignore other information displaying indicators such as the #2 navigation and ADF. Professional pilots will rapid scan all necessary instrumentation to maintain a broad picture of the aircraft's status in relation to the landing field or other intended destination.

Before beginning a major task such as the panel modification covered earlier, the technician or installer should pull the glareshield and some of the instruments to allow for unobstructed viewing of fuselage structure, plumbing, and harnesses relating to potential new work. Sure, there may be a lot of time spent formulating how to approach the modification and forecasting alternatives, but the upside is that hundreds or even thousands of dollars may be saved by preventing costly mistakes during the final phases of the modification. Part of this planning process is the factoring in of the cost necessary for you to perform quality, on-schedule work for your customers. The FBO must be able to finance their operations, so don't undercut profit margins to the point of affecting other customer commitments.

Panel-mounted avionics equipment requires a secure, trouble-free installation to prevent future mechanical failure of the rack/tray mounting system. There are several possible installation methods, but all require that the installer allow for cosmetics, forward "G" force of the plane during landings, vibration during engine operation, constant handling, and clearance for harnesses and plumbing. Vendor installation manuals will illustrate their recommended methods for installation, however this may not satisfy the particular custom installation that may be performed on the next modification.

Customers' requirements should be the first consideration. Will it fill their needs, and will it fit the already restricted real estate behind the panel? Can the shop perform a quality installation that will meet

Radio and instrument panel removed for modification.

FAA and FCC specifications? Can the delivery date be met? Finally, is it profitable to the installer?

Short-changing customers, time after time, will result in them limiting their contact with your facility. First, make sure that you will make a profit, and then commit yourself to meeting the customer's requirements and delivering his or her plane after the modification, on schedule. Take every possible precaution to reduce potential errors, crosscheck the installation requirements against vendor specs, check for structural integrity; in short, use every bit of your background to make the modification a profitable, quality venture.

One of the first factors that should be evaluated in determining if you are making a quality installation is the condition behind the panel. Is there air movement? How safe is the equipment during long periods on a hot ramp? Ambient air temperatures can adversely affect avionics equipment, severely limiting service life. Environmental conditions such as the congestion behind the panel (reduced air movement), venting through the glareshield, and other temperature-radiating factors, such as other equipment and the "greenhouse effect" from the sun entering the windshield, are valid trouble points. Any temperature buildup beyond the recommended operating ranges (TSO'd at 131°F) will start to have degrading effects on the life of the radios at a startling proportional rate. For each 10°C (18°F) rise in temperature, the life of the solid-state circuitry is reduce by 50 percent. Although they might continue to operate for now, when summer arrives, they may fail without warning after sitting on an extremely sunny ramp, or become intermittent during an upcoming flight because of a component that weakened from the excess heat.

Some manufacturers recommend that cooling fans and ducts be added to the radio stack to stabilize operating temperatures. It is also best to leave at least 1/4 inch between the radios to allow airflow. The cost for cooling is not high—much less expensive than the repair bill for failed avionics—and will vastly increase the reliability of the avionics equipment. Another thought is that the customer should be warned that a sun shield or hood should be placed over the windshield while the plane is kept out on the ramp during hot days. This will serve to dramatically reduce the greenhouse effect mentioned earlier by up to 20°F.

A variety of methods are employed in supporting panel-mounted equipment. Some factory installations (before the hardening process) simply fold the extended lip of the panel face back into the equipment cutout, forming an angle. The aluminum is then hardened for structural integrity. This angle is then match-drilled and supplied with nutplates to accommodate the mounting holes in the equipment "can." This method is secure enough, but because of the bend radius, there may be gaps visible around the edge of most of the modern equipment bezels, unless an overlay is used as a cosmetic "coverup."

This writer prefers a simple, clean instrument or radio panel face, without the simulated wood finishes and other "pretties" that some modification installations offer. Epoxy paints offer an easy maintenance surface but are harder to match if any modifications are done. The overlay mentioned earlier in this chapter would work quite well here if the prehardened approach is taken.

Radio panels should be simple, flat metal plates with the appropriate cutouts and fastener holes. Installation, removal, or modification of the panel is vastly simpler when you don't have to deal with radii and varying angles. Access to both sides of the mounting angles for rivet installation allows for clean and quick riveting procedures. Inspection of the completed assembly is also facilitated by the lack of obstructions that can be caused by exotic panel design. While this sounds good, the reality is that you never know what you might run into, and there are a multitude of potential problems and difficulties that are an everyday part of aircraft modifications.

If the modifications were to increase the panel's weight beyond the present capacity of the shock mounts, then the existing shock mounts will have to be replaced with a more rigid design. *Note:* During your panel deletion process, inspect the rubber shock absorber material that makes up the panel shock mounts, because time and airborne chemicals tend to deteriorate the material, thus it loses its resiliency.

Check to see if there's interference with primary airframe structure before you cut into the panel. If interference exists, extensive engineering design modifications may be required in order to maintain structural integrity, unless another more desirable location can be found.

When extensively modifying an existing installation, there are two choices: 1) remove and create a totally new panel, or 2) create an overlay. A metal plate with the appropriate cutout(s) is laid over the old hole and secured with fasteners (nutplates, screws and nuts, or rivets). The problem here is appearance, but this may be offset by the benefits of expediency and ease of installation. The overlay can be painted as required to match or contrast the receiving panel. However, as in any panel modification, the amount of metal removed from the original panel *may* affect the structural integrity of the instrument panel and may require the input of an engineer to diagnose and correct any potential problems. The overlay mentioned previously can also serve as a doubler to restrengthen the potentially weakened panel structure. This procedure was used in many modified Cessna Skymasters and seemed to work fairly well.

Usually a doubler will suffice to strengthen the weakened panel. The doubler will need to be riveted to the backside of the instrument panel and surround the affected area (unless used as an overlay fastened to the panel front). It is important to make sure all aluminum doublers and angles are corrosion-proofed prior to final assembly, including the back of the panel. When installing the mounting angles to the radio tray (also known as "can or rack") with an added doubler, rivet the angles onto the doubler with the rivets passing through the doubler and the face of the panel. Leave out the rivets in the doubler that will accept the equipment mounting angles until you are ready to install the angles. The rivets that secure the angle will pass through the front of the panel, through the doubler, and finally through the angle. Complete the riveting operation by squeezing the rivets to a satisfactory set.

A first step in installing a new panel-mounted box is to obtain the prints or installation drawings for the black box. Dimension definitions will be required but aren't an absolute necessity if the actual black box and mounting systems are readily available. The dangers to using the black box in any modification lie in the potential damage that could be caused by foreign object contamination, such as burrs and metallic dust. Measure the size of the "can" and lay out the dimensions on the face of the radio or instrument panel with an ultra-fine-tipped marking pen. Reference all layout lines to a square side of the panel. Check both the manufacturer's dimensions and the rack measurements once more before scribing the lines to confirm that they match the rack. Once you're sure of the marked location, scribe the layout lines into the metal. This will assure you of permanent reference lines that will not smudge or erase during cutting operations.

At this point, be cautious, because if the installer hasn't previously performed the following sheet metal operations, he or she should not proceed beyond this stage—at least, not until the installer has

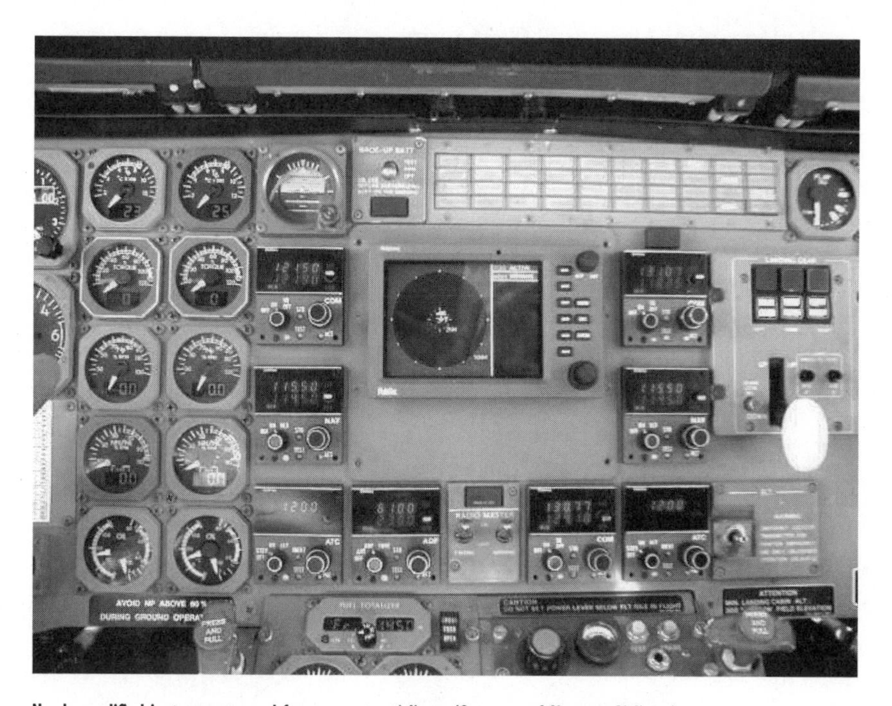

Newly modified instrument panel for commuter airlines. (Courtesy of Skywest Airlines.)

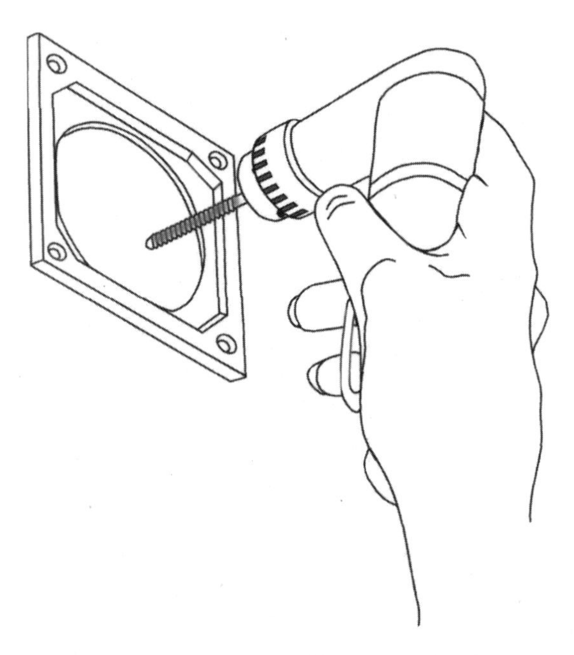

Panel manual routing with Dotco 30,000 rpm pneumatic tool.

practiced routing similar holes in scrap metal sections prior to attacking the real thing! There is only one shot at the panel, and it can become scrap—whereas scrap metal can be played with several times and still remains scrap. Also, practice installing rivets into multi-layered sections of metal before commencing with the riveting operations necessary to complete the installation procedures outlined in the following paragraphs. Also note that the cutting procedures outlined below are the same if installing a new instrument. Using a router template attached to nearby fastener holes, a router block, and a high-speed router, a new indicator hole can be quickly cut. Use the same recommendations for panel-mounted equipment, clean up with files, apply Alodine 1200™*, primer, and paint.

Routing the equipment cutout will require a high-speed router with an RPM rating between 23,000 and 30,000. Dotco of Hichsville, Iowa, manufactures a hand-held router that will serve this purpose very well, as it is very lightweight but still quite powerful. When held firmly, the high RPM will reduce the likelihood of uncontrolled chattering and heat buildup, which could result in severe damage or very rough edges to the panels and the surrounding paint. Minor modifications can be successfully accomplished with the hand-held router and an 1/8-inch router bit when cutting all sizes and shapes of holes, even in the aircraft. The following paragraphs cover primarily a hand-held router, but many of the precautions and methods, such as using a router block, are the same as a pin-routing operation and are to be generically applied.

*Alodine 1200™, a chromic acid, is a registered trademark of Amchem Products, Inc.

A bench-mounted pin router may be selected to perform this cutting operation, but any obstruction of the face of the panel will increase the difficulty for the operator, unless a special spacer or block is provided to raise the panel up to clear the work table. The cutter for the pin router should be a carbide, two-flute, 1/4-inch router bit that is appropriately sharpened for aluminum work.

A router block will be necessary to guide the router along its prescribed route. The manufacturer will usually provide detailed template drawings as part of the installation manual that can be utilized to locate and scribe the intended dimensional shape on the panel face. Take the template drawing to a local machine shop and have them create a steel router guide, representative of the cutout, with a 1/8-inch set back, if you use a non-cutting standoff on the cutter shank. *Warning:* Cross-check the manufacturer's drawings against the actual component before creating a guide template.

Use protective tape applied to the back of your new router block to prevent a burnishing effect on the painted surface of the panel. This approach should reduce the possibility of damage when attempting to cut a hole, without having to repaint the panel. Locate the block on the instrument panel in the prospective area that is to receive the cutout. Clamp the block in place with small "C" clamps, using nearby instrument holes to access the block (block can also be screwed to panel using the indicator mounting holes).

With the router block in place and protective goggles, begin to set the router bit against the inside edge of the template cutout and make a practice run without power on. When ready, make a 1/4-inch pilot hole through the panel with a drill motor (within the cutting area). Activate the router and apply a steady but slow pressure outward against the router block. At the same time, move the cutter clockwise along the peripheral of the block. Keep this pressure up throughout the entire cutting operation—do not force the cutting action—and allow the bit to pull itself through the metal.

Excess pressure during the routing will cause heat buildup, even as far as damaging the paint and causing aluminum gaulding, which will foul or clog the router bit. Watch the cutter as it moves through the metal: Is the drill point starting to glow or beginning to rapidly wear away? A steady flow of crisp metal chips should be flying away from the cutting area without any signs of smoke or excessively rough edges along the prescribed cutting zone. A good operator will always check the ongoing cut and his or her router bits; this will reduce the risk factor and resultant personal injury or panel damage. A vacuum hose held close to the flying chips would catch most of the debris.

During the cutting operation, the router motor should be maintained vertically (approximately 80 to 90°) to the surface of the panel. When approaching the last 1/4-inch of the cut, pull the router bit towards the center of the cutout (within the pilot hole). This will leave a tip of metal extending a short distance into the cutout. The purpose of this maneuver is to prevent a sudden release of the bit, possibly breaking the bit or gouging the cutout edges and router block.

Remove the excess with a smoothing file, followed by the filing and cleanup of the entire edge of the cutout. Work the cutting edge of the file at a maximum of 45° angle to the surface of the panel, allowing it to act as a straight edge across the work, thus reducing the possibility of uneven metal removal. A vixen file is recommended for smoothing the rough edges as it cuts efficiently and leaves a clean, relatively smooth finish. To square the corners, I suggest a "Pillar"-type file with a "0" grade cut, followed with a #2 or 3 to "polish" off and complete the filing.

Prior to completing any assembly steps, all metal surfaces that were filed, routed, or sanded will require corrosion protection applications to the unprotected metal surface. First, clean with a degreasing solvent such as Trichlorethane, MEK, or Alpha 564™. Next, apply Alodine 1200™ (anti-corrosion solution for aluminum), and allow to sit for one to three minutes, then remove with a cloth saturated with water, being careful not to remove the still-soft Alodine 1200™ coating. Install or mate surfaces, as required, within one hour, as Alodine 1200™ becomes non-conductive after the above-mentioned time limit. Original surface finishes must be replaced after the corrosion proofing process is complete using coatings such as primers and paint.

Finally, check the entire installation with a milliohm meter or bonding meter before continuing. A check performed across all bonded and mated points should read less than 3 milliohms, if properly done.

Once the opening or cutout is completed, support angles will have to be installed to attach the avionics can or racks to panel. Suggested material for manufacturing the mounting angle brackets is Spec. Standard AND10134. The dimensions that will work best in most cases is dash number 0601 (.50 in. × .75 in. × .063 in.). The 1/2-inch side is secured to the panel face while the 3/4-inch side is used for securing the avionics can or tray. The larger 3/4-inch surface is used for rack mounting and allows for installation of a variety of nutplates, rotated as is appropriate for the specific installation.

Correct router procedures and good installation techniques of the support angles is critical in maintaining bonding standards and structural integrity. Knowledge of sheet metal practices would be of benefit here. *Note:* No illustrations are provided; this is to ensure that the installer uses vendor-supplied specifications and not uncontrolled recommendations from an outside source.

Securing the angle brackets to the instrument or radio panel can be achieved with either screws or rivets. Consider carefully what future changes or updates may be in the offering. Brackets fastened with screws will allow for easier installation and removal than if they are hard-riveted to the panel. For cosmetic purposes, the rivets will allow for a continuous surface devoid of unsightly screw heads. Of course, screw heads being visible and considered unsightly is a subjective opinion and will only play a major part as part of the cost savings when balanced against the benefits derived from future repairs and modifications.

Whether installing rivets or screws, when a flush fastener is required make sure that a countersink with a cage is used that will allow depth control for all the planned holes. Hold the cage flush against the panel to prevent a misshapen hole, which may leave a gap under the fastener head. Even though a preset countersink cage is used, it's better to be safe than sorry and easier to double-check each hole as you proceed deeper into the panel than to try to put the metal back after going too deep. This is only a precautionary measure and should not be necessary if the setting is locked into the cage, and the countersink has not come loose.

Rivet sizes used to mount the angles in most cases might be AN426 or MS20426 (countersunk). These would penetrate through the panel, from the front, through the angle where it will be squeezed or compressed to hold the angle in place. Again, remember to countersink to slightly less than flush if you wish to shave the head after the riveting process sets them. If cosmetics are not a problem, then a universal AN470 or MS20470 should work quite well. A diameter of 5/32-inch should prove more than sufficient for most all installations. Review the chart provided by your shop for rivet shank length.

A discussion of several riveting techniques and precautions is appropriate at this stage of the modification.

1. Use rivets of the same alloy as the material that you are fastening.
2. Know the thickness of the material to be riveted.
3. For most applications, the diameter of a rivet will be not less than three times the thickness of the material affected.
4. Too large a rivet may deform the material.
5. Too small a rivet may not survive the stresses imposed upon it.
6. Compare the rivets to be used against existing rivets—this will help retain panel conformity.
7. Edge distance from the center of the rivet to the edge of the angle or work piece should not be less than 2 1/2 rivet diameters.
8. Rivet spacing should take into consideration the clearance of the rack fasteners (nutplates and screw shanks).

9. Locate the first rivet within the allowable edge distance from one end of the angle and the second one from the other end. The balance should be equally divided center to center, not to exceed a 10-rivet diameter.
10. Receiving holes for rivets should not be overly tight and all burrs must be removed to allow for correct seating.
11. The same practice that is followed in the composite industry must be followed here: Drill the holes undersized then follow through with the correct size.
12. Try to use a "C" or alligator-type squeeze; they will exert adequate clamping during the set process.

Before installing more than one box, determine if the front bezel of the panel-mounted avionics unit is larger than the cutout. If so, then spacers or shims will have to be temporarily added during the assembly procedure to compensate for the bezel width. Always try to start with the top unit and work down towards the bottom. If you have to remove additional metal to allow for clearance then it would be best to perform the hand fitting at the bottom rather than at the top. This is especially true when attempting to keep newly installed eye-level units in line with other nearby equipment. *Note:* Because of space restrictions, close stacking might be necessary, but additional cooling should be added if none now exists, to prevent future failures.

Several factors must be considered before jumping into what may be a bed of worms. First, you must be aware of the spacing of the angles in relation to the width of the avionics mounting trays or "cans." In many cases, clamping the angles to the can with "C" clamps should suffice to temporarily hold them in place while determining accurate spacing. Once the spacing is determined, clamp the angles to the panel and release the clamps holding the can. The second concern is the depth of the can in relation to the surface of the instrument panel; this is critical in order to obtain a flush fit and secure locking of the radio into the can. If "C" clamps aren't practical, then use temporary spring-loaded mechanical fasteners. Another method is to use self-tapping countersunk screws; the countersink will tend to center in the mounting hole, holding it in place for equipment prefit inspection. (The holes used by the c/s screws are not to be enlarged beyond vendor-suggested mounting screw size.)

We are therefore dealing with three crucial factors that can affect the accomplishment of an well-aligned, flush fit to the instrument/radio panel.

1. Depth of the tray or rack.
2. Correct spacing of the angles.
3. Alignment of the can to the cutout.

Pre-drilling both angles (3-inch angles) with a #40 drill bit at three places in each angle will give you a tie point to Clico™ (spring-loaded clamp) the angles to the panel (temporarily). Once the angles are clamped to the panel (two places on each angle) you can remove the clamps and insert the can and determine the depth location. If you aren't familiar with the type of "can" and its relative position within the cutout, then install the avionics box and lock in place. Keep the box perpendicular to the panel and lightly mark the side and bottom of the rack with a scribe, remove the radio, insert the rack until the scribe lines are 1/16-inch aft of the back side of the panel. This will allow for some tolerance in the locking mechanism.

While the rack is held at the scribe lines, clamp it securely in place and mark with a spring-loaded center punch. The rack can be removed and the marked holes drilled to a #40-drill size. Again place the rack in place and visually inspect to determine if the drilled holes match the rack's mounting holes. If the holes align, then drill out to a size that corresponds to the "can's" mounting holes. If one of the holes is slightly out of alignment, pull that hole using the #40 drill bit, then finish drill the hole to its finished size. Install nutplates and rivet in place (I do not recommend floating nutplates).

Precautions should be taken to not use fasteners with magnetic properties in the area of sensitive instrumentation or any compass-type indicators. One inherently magnetic fastener (steel) should not present any problem, but the total number of fasteners that could be amassed in an instrument panel can be staggering, especially when you consider both nutplates and screws. Stainless steel screws and nutplates are considered to be nonmagnetic and should not present any problems. For indicators, black anodized screws work just fine and are commonly found in may instrument panels.

To protect against the possibility of static discharge or inadequate shielding within the panel components, bonding procedures must be followed during the retrofit; otherwise, RF bonding will end up being part of your rework and cost overrides. It would be considerably less expensive to do it now, during the installation, and not have to perform expensive rework. The old adage "Do it right the first time" still holds true, especially in today's more complicated world.

Electrical bonding straps will be required, if not already in place, at both the shock mounts and at two upper-panel mounting points. This will serve to improve the direct current and low frequency alternating current characteristics, but it is a poor substitute for a RF bond. Having the panel at "0" resistance to ground with at least four ground straps should be sufficient in most installations, but for added insurance, 1-inch by 5-inch beryllium copper straps should be installed across the insulating shock mounts. The one by five ratio to which the straps are cut is tuned to radiate extraneous RF at a different frequency than what the avionics equipment employs. Most panels will have braided straps and may not be adequate for RF related problems and should be replaced with the 1-inch by 5-inch straps.

If the sheet metal assembler doesn't install the avionics radios satisfactorily, the radio could end up hanging by its eye teeth, under the panel, waiting to fall on the pilot's foot or leg, ripping the solder connections from their roots. This hopefully is an exaggeration, but I maintain that a good avionics installation involves all facets of manufacture and assembly and not just the electronics. Though we might be avionics technicians, we mustn't forget the marriage forged long ago between aircraft and the radio (body and brain). Without the radios, in today's environment, the plane may not safely arrive at its

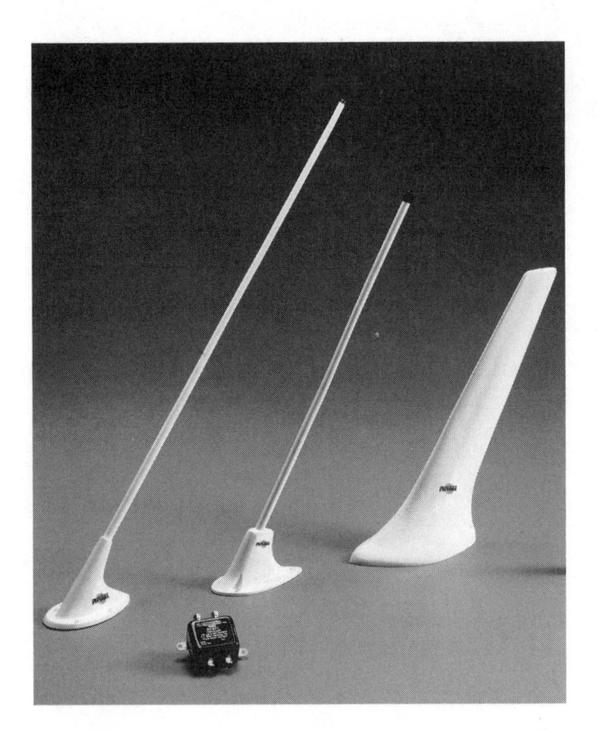

Miller antennas.

intended designation (or on schedule)—and without the plane the radios would not have a purpose.

Antennas

For specific information on antenna-mounting considerations, see Chapter 7 and the respective chapters on the applicable equipment.

Antennas should always be mounted so one person from the outside of the aircraft can remove them. Cost of removing an antenna during maintenance will be much higher if the technician or installer has to wriggle their way into the depths of the fuselage to hold nuts while someone turns a screwdriver. Installing blind fasteners like nutplates during installation will prevent this problem. This is especially important where the antenna mount point is covered by upholstery in the cockpit.

The coaxial cables and their mounting should be designed to allow antennas to be removed from the outside of the fuselage and pulled away from the skin at least 2 inches. Cable clamping should be snug but not restrictive, enough to secure the cable against excessive movement but still allow it to yield to the technician's coaxing during antenna troubleshooting or replacement.

Other Installation Guidelines

There are two primary sources of installation information your avionics shop should be using as technical references for avionics installation. The first is the avionics manufacturer's data, which includes information pertinent to its products. The customer should be interested enough to ask to see this data and confirm that the avionics installer is using it as a reference.

Transponder and DME antennas by Miller.

The second source is broader in scope, but very useful. It is the FAA's AC43, 13-1A and 13-2A. This advisory circular, which is really a pretty hefty book, is entitled *Aircraft Inspection Repair & Alterations* and gives some detailed information on avionics installations and the structural alterations that must be done to accommodate new radios. There is even an entire chapter on antenna installations, and it includes plenty of information about wiring and electrical guidelines. For all the information it contains, AC43.13 is a bargain and is available from most aviation book distributors, pilot supply stores, and many FBOs. As much as the AC43.13 provides, it does not go far enough; therefore, the book you hold in your hands is a valuable tool.

Placarding

Placarding the instrument panel is the final step and can make the difference between a mediocre-looking and a quality installation. Whether installing an entire suite of avionics or a new DME selector switch, it is imperative to identify the switches, lights, and circuit breakers that are an integral part to the installation.

Over the last twenty years, several methods have been used to make placards for instrument panels. The most accepted method, used by factories in their avionics installations, is silk-screening. This is where silk-screen ink is squeegeed onto the panel through a silk-screen template, and it makes for a professional-looking, nearly flawless label. This is also the most expensive and time-consuming process.

Another approach is rub-on wax letters, but these wear off easily after exposure to normal use and sunlight. Sometimes applying clear lacquer on top of the wax letters makes then last longer, but it looks ugly and cheap when not done carefully. Engraving usually looks good, but one little mistake and the whole job has to be done over again, which takes time and even more money.

One simple solution is a self-adhesive film that comes in clear or white and that can be placed in a copier or laser printer to create instant placards. Hawk Corporation distributes this product, among many others. The only drawback is that the lettering is black-on-white or black-on-clear, unless you use a color printer or copier, plus you'll need a computer program to create the placards. Another company, VPC (Vital Presentation Concepts, Inc.), has a photographic process that can create sheets of adhesive labeling for the instrument or panel and other nonavionics applications throughout the aircraft.

The simplest, quickest, and easiest way to create custom placards is using a Kroy or Brother labeling machine. These are available fairly inexpensively from any office-supply store and have a variety of features that makes them ideal for custom placarding, including vertical lettering, various type (font) sizes, and an assortment of colors. The lettering resists fading from ultraviolet light. The actual lettering is on the opposite side from the exposed surface and is virtually impervious to damage.

The surface is such that the installer can remove the glossy finish with a 3M Scotchbrite™ pad or 400-grit sandpaper, leaving a flat finish. With the heavy traffic from pilots' rubbing against the instrument panels and the damaging effects of sunlight, placards are subject to a lot of wear and tear. The labels produced by these units show every indication they can handle the job. Brother has yet another tape label that produces lettering that can be released from the tape back to the panel. The finished product looks like it was screened on, very professionally applied.

Applying Placards

Applying placards properly is an art. The goal is satisfactory appearance and long-term endurance. The following installation discussion describes creating a placard with the Brother labeler.

When dealing with a smooth surface, first simply clean with a white cotton cloth saturated with a mild soap-and-water solution. Using alcohol or other solvents could remove old paint, change its color, or make the cleaned area glossier than the rest of the instrument panel. The soap-and-water mixture is less apt to cause this kind of problem. Rinse with clean water or a clean wet cloth before installing the placard, and allow to dry.

After cutting and removing the placard from the labeler, cut off excess material as close as possible to the letters with a sharp razor. With the tip of your cutting tool, lift the letters from the backing material. When removing the backing and handling the placard, do not touch the adhesive; the oils in your fingers will prevent permanent adhesion to the desired surface. Hold the placard using the razorblade and, bringing the placard into position, let one small corner of it touch the panel, allowing the adhesive to grab. Adjust the opposite side so it's straight, then press firmly on the placard with a finger to seal it in place.

On large placards, use the discarded glossy backing to rub the placard to remove air bubbles. If there are any bubbles left, use a sharp needle to prick the bubbles then rub the placard to force the trapped air through the pinhole.

A textured surface is a little more complicated but can accept placards with a little additional effort. On some lightly textured surfaces such as painted panels, gently rub the area with an ink eraser to smooth out the majority of ridges and valleys, allowing the placard to seal firmly on the panel. If the surface is plastic, such as a pedestal or sidewall or false instrument panel, a small amount of lacquer thinner dabbed on the end of a sharpened but slightly blunted ink eraser will smooth the plastic. Be careful not to smooth out beyond the area that will be covered by the placard. Once dry, install the placard.

If you're concerned about the placard's edges lifting from heat or wear and tear, gently apply a thin coat of flat clear lacquer (not lacquer thinner) on the top of the placard and carefully around the edges. You can get a small quantity of lacquer by simply spraying a small amount into the cap of the spray can. Allow it to evaporate a little because it is a lot easier to handle when slightly thicker. The discarded backing tape serves quite well to transfer a thin quantity of clear lacquer to the edges of the placard. The sharp edge of the placard will wick the thin lacquer along its perimeter. Allow the lacquer to dry thoroughly before touching the placard.

Vinyl surfaces such as those used on Beechcraft panels are nothing more than an industrial-grade, vinyl contact paper. This is a petroleum-based product that excretes an oil film that will resist adhesion of the placard. Simply clean the intended location with soap and water or alcohol, let dry, and apply the placard. Again, the lacquer application will help assure the placard will stay in place.

One other use for the labeler is to make placards for annunciator lights, circuit breakers, and for wire identification. To make a wire placard, simply print out the desired number twice, leaving 1/4-inch between the two numbers. Then remove the backing and wrap the label around the wire and stick the two free ends to each other, leaving a clearly tagged wire, which will do wonders for making maintenance easier. Yet another use is to identify terminal blocks, Cannon plugs, and other connectors.

Soldering: General Application

Two basic types of soldering are used in aviation: one is lead-alloy soldering; the second is silver soldering. Both are used in wiring termination and mechanical connections. Silver is predominately used for higher-stress or -temperature conditions, while lead is used for electrical and electronic systems.

Lead-alloy soldering of delicate assemblies requires more detailed application of your skills as very little solder is used per contact and the heat can easily damage the very job you wish to repair.

There are three very basic steps to effective, quality soldering.

1. Select an iron sufficient to perform the work, but not so large as to overheat and damage the surrounding connector structure.
2. The tip of the iron must be clean and pre-tinned with solder, excess wiped off on a damp organic sponge, which are becoming harder and harder to find with all the synthetics on the market. Silver-coated tips are easier to maintain, but will corrode under the coating and deteriorate without showing externally.
3. The work, connector pins/sockets, must be clean and pre-tinned, prior to assembly.

A 25-watt, high-grade, iron-plated-tip soldering iron is normally adequate for most contacts encountered in radio installations.

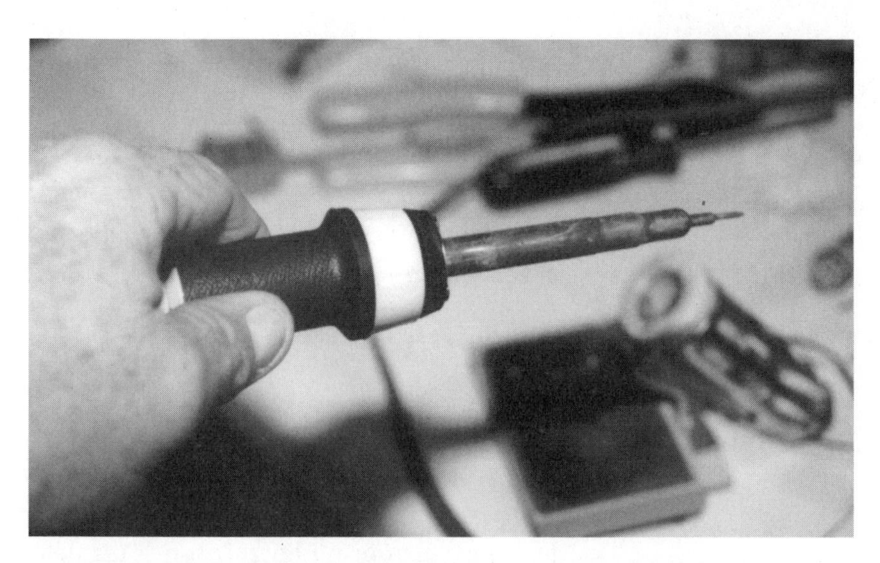

25-watt soldering iron and related tools. Note the damp sponge and protective shield on the iron station. Each time the iron is to be used, the tip is cleaned on the sponge. Use only natural sponge; do not use synthetic sponges. Apply a new coat of fresh solder to protect the tip from oxidation until it is needed again.

Occasionally, a temperature-controlled iron or a heavier unit might be needed for electrical repairs. However, *do not* use the "gun" trigger-activated style unless you have plenty of room and the tip is appropriately designed for the job. These are difficult to maneuver in high-density connectors and the heat is usually well over 100 watts. Besides, in solid state circuits they aren't recommended. Improperly grounded irons will also damage electronics. For most other aggressive applications, a hand-held, permanent-on, 60-watt iron would do nicely.

There are different styles, but a 90° handle grip would be easier to maneuver when sitting down at a workbench or in restrictive spaces. If 115 VAC isn't available to you, there are butane-torch styles (*not to be used inside the aircraft*) and 12 volt D.C. operated irons that'll do quite nicely, even on electrical and radio repairs. If you really want a good portable iron, check out the Wahl unit, distributed by Hawkins Associates Company, Inc. Wahl makes a 900°F iron, with an Iso-Tip® that uses two AA-replaceable cells (batteries) that can be unloaded and a fresh set installed on the fly. There is even a storage area at the rear of the iron for spare tips. Just when it seemingly stops amazing you, you find it has one more little feature that makes this iron indispensable. Wahl designed the iron with a protective cover for the Iso-Tip®, which protects when laying on the bench. This iron works exceedingly well when the work is being done in the plane far from the hanger and an extension cord. The cost for this unit is very reasonable.

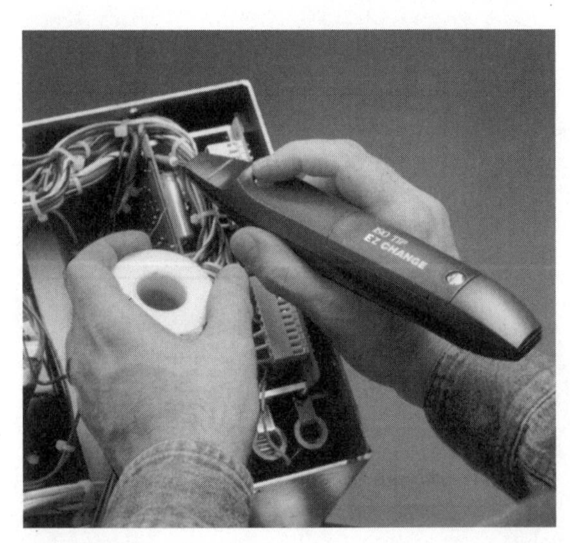

900°F soldering iron by Wahl with replaceable, off-the-shelf rechargable-battery cells. This also sports a storage space in the rear of the iron, plus a tip protector for the Iso-Tip.

Something that is often overlooked is the length of the tip from the body of the iron. There will be an approximate drop of 100°F for every inch of tip on a 1/4-inch diameter shaft. Make sure the tip is appropriate and adequate for the task at hand. The surface of the tip making contact with the work must cover as much area as possible to transfer the heat in as little time as possible. Too much heat can damage the subcomponent or the component, or both. The tip should be a copper core plated with high-grade iron, with a stainless steel shank. Nickel/chrome should be coating the area aft of the tip and just before the tip body (section that fits into the soldering iron body). The stainless steel prevents seizing in the soldering iron body. The high-grade iron-plated tip allows for easier maintenance.

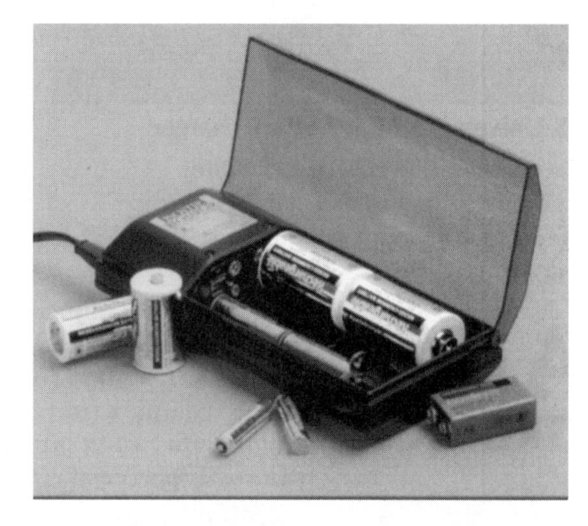

Recharging station for Wahl battery cells. Also accommodates many other sizes of cells such as the AA, AAA, and 9 volt.

Hand tools listed below are needed to terminate the conductors, strip off the insulation, hold the wire during soldering, form the hooks or bends, and solder the wire.

1. Aviation-style strippers.
2. Diagonal electronic-cutting pliers.
3. Electronic needle-nose pliers (3 1/2-inch).
4. Solder vise or clamp method to support work during assembly.
5. Round-nose pliers (3 1/2-inch).
6. 25- to 60-watt soldering iron.

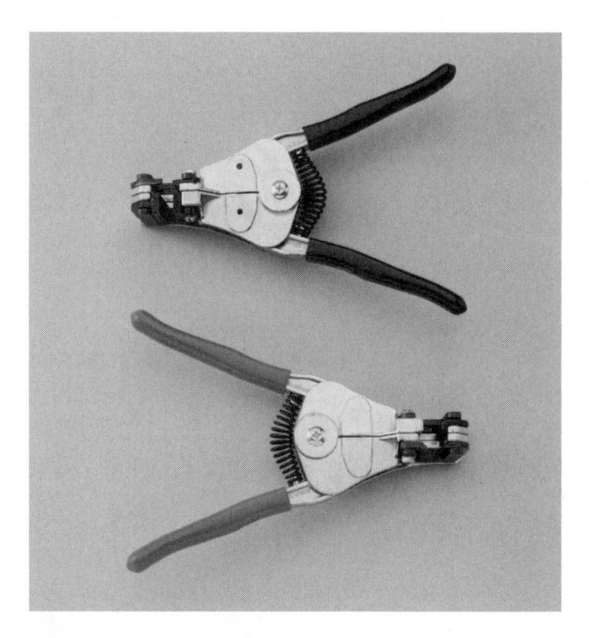

Teflon and standard blade strippers.

Don't forget the solder; use a mildly activated 60/40 for all electronic and electrical work. Other, more highly activated types will corrode connections and cause future failures, particularly on circuit boards. To prevent excess solder from flowing into the

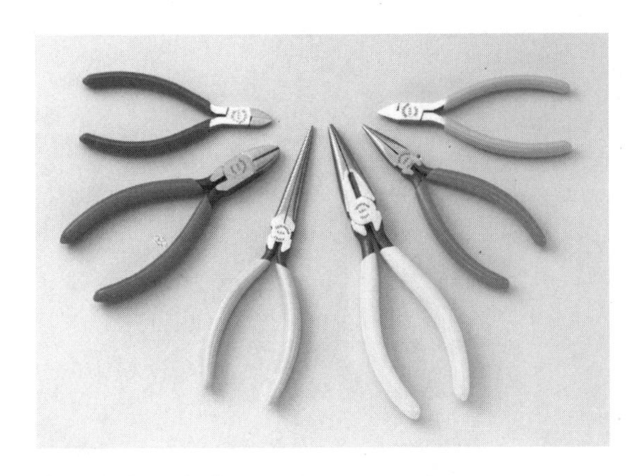

Electronic pliers selection.

affected joint, select a smaller diameter. There are different sizes of solder, depending on the work being performed. Another factor is the rosin; some solders are designed with up to four cores. Select according to your individual requirements.

Practice, and plenty of it, should be done before attempting to solder any electrical or electronic connectors. Some miscellaneous parts may be needed for practice samples and can be cannibalized from discarded connectors through disassembly or be simulated with the use of solderless terminals. To use a ring-type terminal as a substitute, simply remove the plastic insulation from closed-loop-type terminals and perform your trial soldering on the ring, often called a closed loop. Radio Shack is a good source for miscellaneous terminals, connector, pins, and sockets that can be used for practice.

Conveniently place all of the above-mentioned tools out on the workspace. Before starting the process, confirm that everything needed is quickly available. The contact or terminal should be secured in the rubber-protected jaws of the vise, in a manner that allows easy access during conductor insertion and soldering iron application.

One serious problem with soldering iron tips is the amount of heat that is held in reserve during the mating of the tip to the work. If the tip is too small, the heat dissipates too quickly and then requires longer dwell time. This usually results in damage to circuitry or the board on PC boards. Temperature at the point of contact should be about 100°F above melting point of the solder.

Below are some guidelines for soldering iron-tip dwell time:

1. Circuit boards should not exceed 1/4-second contact dwell; tip temperature about 700°F.
2. Mounted bifurcated terminals should not exceed 1 1/2-second dwell time; tip temperature 800 to 850°F.

3. Free air connector contacts should have a maximum dwell time of 1 3/4 seconds; tip temperature 810°F.

Because of the danger to circuit boards and their components, the product Chip Quik is the only choice if lower cost, easy removal is desired. Check out http://www.hawkinsassoc.com for more information.

The insulation must be removed from the end of the wire before any soldering begins (varies depending on diameter of connection or depth of contact). The lay of the wire must be maintained to allow proper insertion into the solder cup. Appearance should be uniform with parallel strands of wire from the insulation to the removed end of the conductor. If, after stripping with the correct strip die, you find that the individual strands of the conductor have been damaged by nicking from the stripper blades, you must refrain from further use of the strippers until the tool has been repaired or adjusted. Even minute nicks can cause the wire to break at a later date. Lay the stripped wire off to the side for the moment and continue on to the next step.

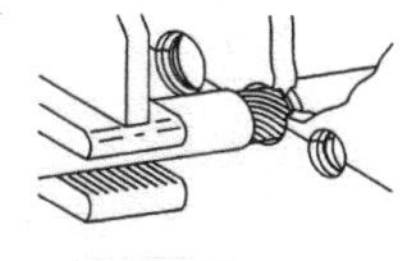

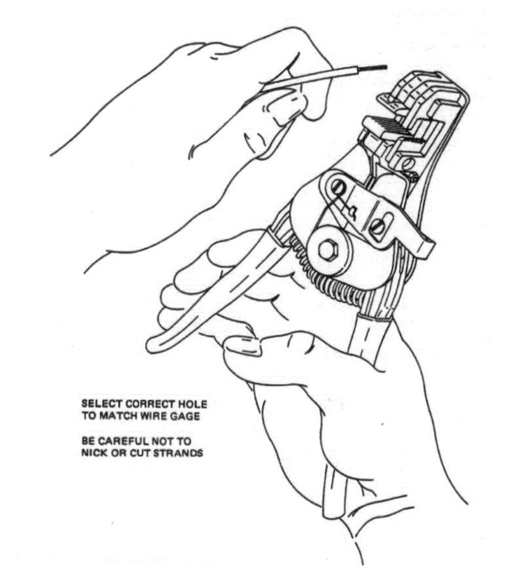

Stripping insulation from wires. Following the correct procedure will result in a clean, neat strip with all strands laying with the proper twist.

When the tip of the soldering iron reaches a temperature that melts the solder (700°F), apply the end of a piece of solder to the tip and rub in a small circular pattern until it begins to melt. This is known as tinning and should transfer a small quantity of solder to the working tip, which will prevent oxidation during the soldering operation. This process should be attempted immediately as the tip reaches the melting point; otherwise the tip will begin to oxidize, making tinning difficult. Another method is wrapping a small quantity of solder around the tip before it heats up. As the tip reaches approximately 700°F, the solder will melt, flowing around the prepared surface. This procedure should be repeated throughout any extended job. If the tip is excessively corroded from previous use or the iron plating has eroded, leaving only the corroded copper, a smooth file should be used to "dress" the tip to a pyramid or conical shape, depending on the work to be performed.

As much of the tip of the iron should make contact with the connector terminal, resulting in the maximum amount of heat transfer possible. Failure to properly apply the tip to the work can result in excessive heat reaching the surrounding connector body, causing damage. Additionally, when the solder does start to flow, it might wick clear through to the opposite side. If a problem arises in obtaining good heat flow, try moving the tip in a small pivotal rocking motion against the terminal, until solder flows freely.

Because capillary action is the medium through which liquid solder flows while joining two metals (similar or dissimilar) it is critical that the surfaces be thoroughly cleaned of burrs and contamination. This can be accomplished with liquid or abrasive removal, depending on the application. Liquid solvents work better, for example, on hook terminals, while abrasive (braided shield stock from Radio Shack) removal serves to remove corrosion from resistors, diodes, capacitors, etc., prior to installation. It often isn't necessary to clean the conductor of an insulated wire as the corrosion factor is much less and the rosin or flux in the solder will perform that function as the solder flows in and around the connection.

The inherent strength of a soldered joint is derived from the wetting action that occurs when the filler metal melts and flows throughout the joint area. The molten liquid metal dissolves and reacts chemically with the surface layers of the base metals, forming a thin layer called the joint interface. To enhance this action, rosin or flux within the solder removes and prevents reformation of surface oxides on the base metal. This is mandatory in production lines where any inhibiting factors will slow production. Applied just prior and during the heating action, the flux removes the surface oxide barrier that prevents the wetting action upon which successful soldering depends.

Pre-tinning of the stripped wire will be the initial test of your iron's ability to properly heat and melt the solder into the individual strands of the prepared wire. Place the tip of the iron 1/8-inch from the insulation and apply the solder at the furthest point from the iron's tip. Predictably, the solder will flow toward the heat source, thoroughly saturating the individual strands with liquid solder, forming a homogeneous mass. It's important to immediately remove the iron's tip before the solder is drawn or wicked up beyond the desired 1/32-inch. Solder flow or wicking beyond this point will form a potential stress area of inflexibility. The strands could conceivably start to break, one at time, until the wire falls loose, causing a failure.

Before starting each soldering application, reclean the tip of the iron; lightly brush the freshly cleaned and shiny surface with a new coating of solder. This will establish a new, non-oxidized surface that will make better contact and greater transfer of heat during each series of soldering operations.

The iron should always be placed opposite to the solder application. The melting solder alloy will move toward the heat source, joining both the wire and the contact into one unitized, strong electrical connection. If you need to get the solder to flow to a stubborn point, move the solder and perform a gentle, circular rubbing action without lifting off the contact; this may break the heat transfer contact and prevent the solder from flowing to a more distant area of the joint. Each completed solder joint should have a shiny, clean filet of solder between each strand of wire and the cup or terminal.

When each connection is completed, inspect for correct insulation gap, wicking, excessive solder, pitting, chalk-like appearance and a good filet around each strand, without voids. If you leave your iron on the terminal, three things may take place: one, oxidation of the solder; two, overheating of the surrounding connector; and three, wicking of the solder up the strands under the insulation. This will stiffen the wire, making it more susceptible to breaking.

Circuit Boards and Wiring Repairs

Circuit Board SMD Removal

Once circuit boards are released from stock or final production to be installed into final assemblies, numerous problems can occur that will require intervention to return them to like-new condition. Repair is not as simple as crimping a terminal onto a wire; either overheating or mishandling can damage the delicate circuits that compose PC boards. The components are highly sensitive to not only heat, but also electrostatic discharge (ESD). Approximately 90 percent of all ESD damage is from "unaware" individuals handling the boards without using adequate precautions. Those responsible for either repairing or inspecting printed circuit boards should have an overview understanding of the production processes. It is also necessary to have extensive experience in all the possible known errors that could arise.

Removal of surface mount devices (SMDs) is extremely difficult and usually ends up with the board finding its way into the trash

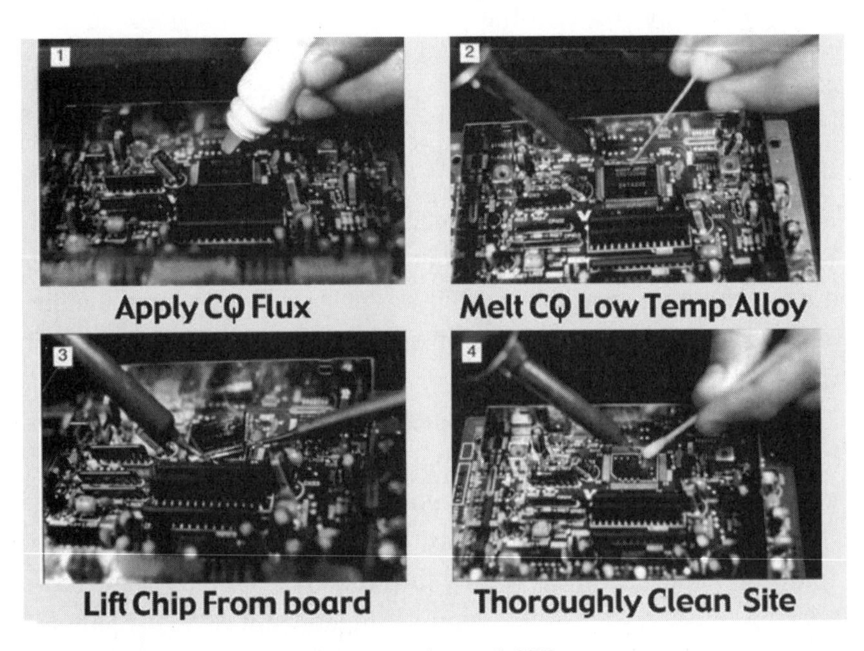

Chip Quik removal process; the advanced answer to safe removal of SMDs.

next to the workbench. At this time, unless the repair facility is in the market for spending thousands of dollars on reflow machines, there is only one way that will allow the repairperson to remove the device and not damage it.

Chip Quik Desoldering Station; this makes removing SMDs so simple a ten-year-old child can easily do it.

A new patented process called Chip Quik™ that takes from the bronze age is now applied to the evolving electronics field of SMDs. It is now available to those interested in removing those hard-to-remove chips without damaging the chip or board. It is a process that uses a little-known principle of physics involving two metals of dissimilar melting temperatures. When a metal of lower melting temperature is heated along with one of a higher melting temperature, the higher melting point metal will actually melt between the lowest melting point and the highest. The resulting melting temperature of the higher melting metal is now under 300°F. Why is this new application so important? The answer is quite simple. What took thousands of dollars to assure safe removal of an SMD can now be done for under forty dollars and the repairperson is not forced to do the repair in a specialized area. Solder requires over 600°F to melt and flow freely, but with Chip Quik™, solder will melt at approximately 300°F, or less. Chip Quik™ Inc. developed this process for the nonavionics market, but it is steadily creeping into the general aviation community and has already met the challenge in the commercial aviation market. Contact Hawkins Associates Company, Inc. at (800) 433-2612 for more information on this unique product.

Electrostatic Damage

I know of no easy way to say this: The handling of circuit boards has become very sloppy. Many technicians actually believe that there is no damage done by handling them with your hands and walking from stock to the assembly area. Wrong!! The odds of having inflicted permanent damage to a circuit board are quite likely. The voltages developed from walking across a carpet or waxed floor can be thousands of volts. If you have ever been shocked by someone who walked up to you and touched you, you know what I am talking about. There are three stages of damage from ESD; one is total or catastrophic failure to the circuit, the second is delayed failure; it will fail at a later date. The third and more insidious failure is circuit degradation. The board seems to operate properly when installed, but will cause erratic operation that is difficult to locate.

In 600 B.C., one of the earliest "technicians" observed the electrostatic charging of objects. This man, Thales of Milet, logged in his observations and was able to pigeonhole his discovery into the annals of history. What we all wish he had done besides discovering this phenomena was to get rid of it. To this day, electrostatic charging is still the reason for many avionics glitches. These observations were based on only that which he could see and "prove" in that time and place. ESD is not visible with the naked eye, yet today, even elementary school children are aware of its effects and how much fun they can have shocking their parents. Aviation professionals should be just as familiar with the phenomena and its potential dangers.

Electrostatic discharge or ESD is perhaps one of the most underrated problems that occur every day within the electronics industry. This includes everything from television, toys, and computers to the aerospace industry. Static electricity can damage circuit cards or boards while leaving no immediate, outward sign of injury. This may come as a surprise to those who haphazardly install memory or other upgrades to computers. Sure, you may have handled circuit boards and solid-state devices for years and never had a problem. However, there is a "but" here; the trouble is, how do you know? Much of this damage may not surface for days, months, or even years. Static damage is cumulative, just like solar damage to human skin. Each time the static charge hits the discreet circuit, it causes a little damage, but not enough to totally open or short the circuit. Finally, after numerous shocks such as vibration, heat, and being turned on and off repeatedly, the circuit catastrophically fails. At that time, the reason for the failure is almost impossible to trace and correct. The diagnosis will either be "random failure" or "electrically overstressed." That is why so much effort is given to prevention, as opposed to correction.

Sure, there are protective measures integrated into the design of electrostatically sensitive devices, but factors such as the speed of the charge, duration, and number of hits can, and have, overrun the devices. Zener diodes are one such device; they just aren't fast enough. Since the late 1970s, newer devices were made available, but they aren't found in older equipment. When people start to make excuses like, "I've picked them up hundreds of times," ask them if would they take that chance with $250 (or more) of their own money?

Ways to reduce potential of ESD damage while working on electronic components are:

1. Turn off equipment; wait for at least ten minutes.
2. Find and touch metal chassis that is connected to power ground return.
3. Attach ground strap.
4. Do not move around after grounding yourself.
5. Touch only the components to be installed or removed.
6. Do not make contact with any of the open-end mating contacts.
7. Lay both the removed and to be installed boards on clean, grounded surface.

Keeping a presence of mind while working around static-sensitive devices is not any different from adhering to any other quality installation standard. Static electricity (over 12,000 volts) from walking on carpet is annoying to us in our homes but can cause damage to solid-state devices on printed circuit boards. The tremendous levels of high-voltage static electricity that lurk in shop areas on the walls, tables, or flooring will either cause permanent damage, change the operating characteristics, or degrade the long-term permanence and quality of the ICs and other electronic components used in the sophisticated avionics systems. As mentioned earlier, the degradation may not show itself immediately but may cause the failure of your radio when it is most needed. When handling any electronic components be careful not to touch the connector pins or any other exposed connections that could be a direct path to the sensitive and fragile integrated circuits that constitute the brains of the radio.

Humidity, type of flooring, shoes, and clothing are all factors that contribute to circuit damage. Prevention methods are the bagging of electronics in antistatic wrap or bags. The components are shipped to customers where they expect the recipients to observe the instructions and warnings. Sadly, people don't. Simply putting on a grounded static wrist strap reduces charges to virtually zero. The reasons for not doing so are the same as those who refuse to wear seat belts.

It might help to imagine the electrostatic field around your body as you move across carpet or other static-creating materials like the ripples formed by a stone dropped into a calm pond. The stone is you, and the field contours your body, extending out in all directions. Any time that field comes in contact with other materials, the potential exists for the charge to discharge into that material.

Touching the metal case of the computer, avionics radio chassis, or the wall receptacle metal box for up to 20 seconds provides vastly more protection than not using the strap but is not considered to be an industry standard. While sitting or standing, don't move! Any motion may create a static charge on a person's body. Grasp the circuit card, being careful not drop it or hold it on any of the circuit tracks. All of these are just precautionary methods to prevent what is more easily stopped dead in its tracks by using static wrist bands, grounded work benches, antistatic clothing, and other preventive measures. The wrist bands are the most effective, especially when new components are brought to the bench. If the work area is grounded, the bag will be brought to the same potential and the danger eliminated. Believe it or not, even if grounded to the bench, if an individual should pull a section of transparent tape from a roll, a charge of static electricity is created and will damage the components.

The amount of charge to cause damage can be as low as 100 volts. Dual In-line Packages or DIPs and other discreet electronic components such as C-Mos may be damaged with as low as 100 volts and as high as 200 volts. Other subcomponents such as the Mos-Fets, will take higher voltages, about 240 to 2000 volts before becoming damaged. Bipolar transistors have exhibited higher voltage failure points of 375 to 7200 volts. Now, remember those figures. The static voltage levels accumulated from walking across carpet is over 12,000 volts; sometimes many times higher, even up to almost 40,000 volts. Tile, especially with some waxes, will produce a static charge on an individual up to 12,500 volts. Even just sitting at a work bench, with no grounding strap, a static charge can develop, anywhere from 450 volts to as high as 3500 volts. If you rub synthetic pants or skirt fabric, an enormous charge is produced. This is why only cotton should be worn while working with electrostatically sensitive devices.

This is one for those supervisors who want the place kept clean and neat. When a charge is accumulated on someone's body, the static charge is constantly looking for a way to equalize. As soon as that opportunity arises, the charge will jump, and if a device is in the way, it is destroyed or damaged. One point that should be made is how a work bench is kept. The more cluttered, the higher the potential for a product that is nonconductive to be present and give up its charge at the first opportunity. Charges on nonconductive materials can remain for several weeks, even months. If drinking coffee is a daily habit, don't bring the Styrofoam cup to the work bench; it carries a static charge. Lunch bags, hair clips, pocket protectors, purses, books, clipboards, candy wrappers, and even eyeglasses are carriers of static. Keep them away from your work area.

So, what creates the initial static charge? Essentially, this surface condition occurs when two materials composed of nonconductive materials are separated. The speed and conductivity of the materials will determine the degree of charge. This can take place in the form of walking across carpet, pulling two materials apart, or rubbing your hands on your clothes. Metals are more conductive; therefore a static charge will spread evenly across the material and quickly dissipate. Nonconductive materials such as plastics will hold charges in pockets, varying across the material. For this reason, the charge remains longer. This is why nonconductive materials cannot be grounded. The charge will remain at other points on the surface. A piece of Scotch tape can be as destructive as the hammer on a gun. Pulling a piece of tape from a dispenser in a flour mill may very well be the last thing you ever do, for a tremendous explosion will most likely be the result. Flour is potentially explosive, and static electricity has blown mills sky high. Think of the damage to an electronic circuit!

Besides ESD, other factors contribute to circuit board failure. Some attention must be given to this issue so not every effort is misdirected solely on ESD. Older studies laid blame entirely on ESD, but other factors can contribute to field failure. Temperature can cause failures by the same process that interstate highways and other long-run materials expand and contract. Within the chassis of a radio, the temperatures rise well over 100°F. As the circuits expand with the heat, an open circuit may occur. As the radio cools, the connection is again intact. This would be construed to be an intermittent failure. Maybe the failure isn't total, just the overall performance is affected.

If the #1 navigation radio system fails, the backup or #2 radio can be used for navigation. If the communications radio quits, the same is true; however, this usually not true for the DME, transponder, or marker. This is why it is very important to make all the correct decisions and do the very best job possible. You can't get out and walk to the next service station when at fifteen thousand feet.

Contaminants are the second main reason for circuit failure. These can be metal powders or conductive (ionic) liquids. Metal being conductive, the failure is obvious, but for the second variety of contaminants, the problem is far more insidious. Sulfur dioxide, ammonia, alkaline mixtures, corrosive soaps, hydroxides, chlorine, oil, caustic cleaners, alcohol, and resins are just some of the corrosive materials that can actively damage circuit boards. Some attack more slowly than others, but they all are potentially damaging.

Electrical fields are yet another contaminant, albeit not visible to the naked eye. Examples are X-rays, cosmic rays, arc welders, conveyer belts, tesla type fields, or plasma flux. Magnetic fields fill out this covey of destructive contaminants. These are found almost everywhere within the factory environment and within the aircraft.

Junction breakdown is the point where circuits meet. This can be a solder joint, mechanically overstressed or even a defective component. Vibration also plays a part in junction breakdown; that is why instrument panels and equipment are found to be shock mounted. Improvement in manufacturing methodology has allowed companies to produce boards that are very resistant to vibration. This equipment can be mounted directly to a shelf. Flaws or built-in failures are something our industry has lived with but should not and must not be a prevalent problem.

Those responsible for either repairing or inspecting printed circuit boards should have an overview understanding of the production processes. It is also necessary to have extensive experience in all the possible known errors that could arise. It is difficult to find many of the defects that plague manufacturers and end users when inspectors don't have the wisdom that comes from either experience or knowledge. An example would be a production defect that is hidden under the conformal coating, or too small to be seen with the naked eye. Having a broad knowledge platform will assure greater quality success in completing repairs.

When inspecting for defects on a wide variety of PC boards, it is best to examine for obvious defects, placing those boards into a different pool than those with less evident problems. As the boards undergo the repair process, they may change their level of failure from the obvious to the less apparent. Defects on boards may be divided into three classes: critical, major, and minor. At all points along the refurbishing process, the original squawks should remain with the PCB (printed circuit board) until it is stocked. Then, those records should be filed for future reference.

To sum it up, conservation is the name of the game. Applying extensive grounding methods and preventive habits are key factors in reducing subcomponent failure and, resultantly, avionics failure. So what have we learned? Use grounded floor mats, antistatic waxes on tile floors, grounded work benches, wrist straps, and keep the work area free of clutter such as candy bar wrappers or sandwich bags. Additionally, implement the use of antistatic carriers such as trays and bags, plus the application of shunts or mechanical jumpers across electrostatically sensitive leads and contacts. An antistatic blower should be added to further to enhance the precautionary approach, otherwise the very air can become electrostatically charged. One more thing: Use nonconductive service tools. Conductive tools can accumulate static charges.

Just as static can build up to over 15,000 volts on the human body after walking on carpet, the same is true for an aircraft, except voltages can exceed 350,000 volts. Slipping through dust or moisture-laden atmosphere is a good comparison to walking across carpet. Static charges build up to dangerously high levels and can cause serious and extensive damage to personnel or ground equipment if static charges are not dissipated on landing. Both helicopters and fixed-wing aircraft are subject to this condition. If these charges should reach the sensitive electronics within the radios, they will be damaged to some extent. RF and electrical bonding is necessary to prevent or reduce this threat. A wrist strap just won't do it. This is the worst-case scenario related to electrostatic charges, and although it doesn't seem like it, the solution is essentially the same. Apply extensive grounding and bonding procedures, and keep the static-sensitive components away from static-laden charges.

Fighting ESD requires looking at the whole picture, not just some of the brush strokes. Electrostatic charges have been with us for a long time, and they aren't about to go away now. The battle rages on with many of the old approaches being put into play and even some of the new. Most static-sensitive devices are, for the most part, a serious target only during the manufacturing process. Don't, however, allow this false sense of security to lull you into ignoring the preventive measures outlined here. The end circuit connections still remain a path for ESD. Electrostatic discharge countermeasures must be practiced as a normal part of doing business with a full understanding of this invisible enemy. Remember, you may be a carrier.

Wiring

Many contacts are still soldered into connectors, relays, switches, and lamps. Performing a solder operation isn't for those who learned with a Radio Shack weekend special. In fact, aviation avionics specialists in the factory environment will spend from 40 to 80 hours preparing for that day they can start to assemble connectors and harnesses. A little secret is the same Chip Quik process for circuit board repairs can be used to remove wires from connectors.

There will be no attempt to teach readers to learn how to solder from this book. It isn't possible; there is no way to minimize the training necessary to properly solder electrical or electronic components.

Tools

Taking a hammer and a screwdriver to an overly tight connector or using slip joint pliers to loosen a nut can hardly be classified as using the right tools. How about this, crimping terminal lugs with a pair of vise grips? Doesn't happen? Wish that wasn't the case, but it happens far too much to be an aberration. At the FBO, on the ramp, or under the protection of the hangar roof the situation is controlled, so there isn't any legitimate excuse for not using the correct tools. It would be different if there wasn't a choice. Being facetious to make a point, an emergency landing in the middle of the desert, with crazed terrorists surrounding the plane, and armed only with a minimum of tools, you do what it takes to get the plane, and its crew and passengers back in the air as safely as possible.

Removing an apex screw with a small common jeweler's screwdriver is like using a wood drill bit to drill through aluminum. It may give the appearance of working, but almost immediately the metal begins to bunch up around the cutting edges, and the bit quickly overheats which is not unlike the screwdriver where the tip strips and the operator overheats. It soon becomes crystal clear that more instruction is needed for those given the job of working on electrical or avionics systems. Several vendors have tools that will serve adequately for aviation electrical and avionics assembly, installation, and repair. The problem is receiving adequate training on proper use in varying situations. Choosing the right tools is a matter of need, personal choice, and vendor preference. Regardless of specific guidelines provided by manufacturers, some tools work better for one person than for another. Several are designed to do the same job, but a few are ergonomically designed.

Several companies, such as Jensen and AMP Corporation, provide tools specifically related to the aviation market. AMP manufacturers everything from coaxial crimpers to insertion tools. Other companies, such as Jensen and EDMO, distribute a wide variety of tools for everything from riveting to soldering. In several cases, however, it may be better to be very selective in the type of tool being purchased. For example, precision circuit board manufacturing or repair

Daniels Manufacturing Corporation screw-machined miniature crimpers.

requires precision heat-controlled irons. Companies such as Hexagon might be a better choice when needing soldering equipment for factory production.

Keeping all the tools readily available is a chore in and of itself. Some shops like Skywest Airlines have taken an approach that not only provides storage, but also quickly displays the missing tools. A roll-around toolbox lined with a semi-hard foam was cut to fit the technician's tools. Each tool has its own home so a tool out of place can be quickly identified. Leaving tools behind in the plane after a job is complete is now in the past.

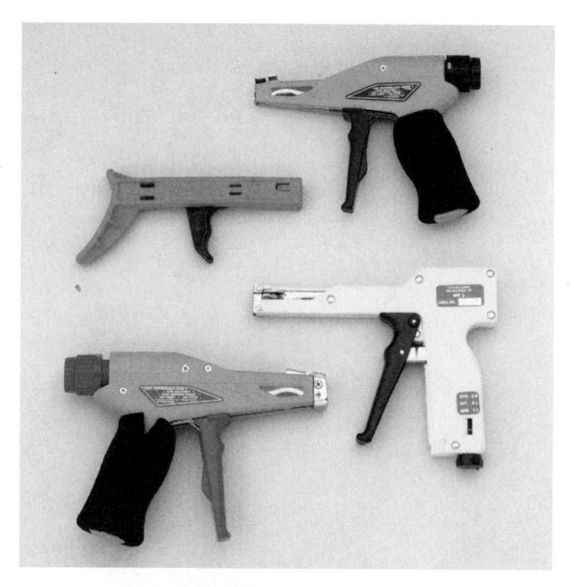

Nylon tie-tensioning tools by Tyton® Corporation.

The Crimping Concept (Courtesy of DMC–Daniels Manufacturing Corp.)

Crimping is a method of firmly attaching a terminal or contact end to an electrical conductor by pressure forming or reshaping a metal barrel, together with the conductor. The forming of a satisfactory crimp depends on the correct combination of conductor, crimp barrel, and tool.

When applied with a properly matched tool, a union would be established which has both good electrical and mechanical characteristics. The tool, will provide these requirements consistently and reliably with repeatability assured by quality cycle controlled tooling. There are several common configurations of a crimped joint.

The electrical resistance of a properly designed and controlled crimp joint should be equal to, or less than, the resistance of an equal section of wire. Specifications state the requirements in terms of millivolt drop at a designated current.

The mechanical strength of a crimped joint and hence its pull-out force (tensile strength), varies with the deformation applied.

The dies in the tool determine the completed crimp configuration, which is generally an element of contact and/or connector design. Some of the design considerations are: a) the type of contact,

Roll-around mechanics tool-box lined with glove-fitting tool depressions.

its size, shape, material, and function, b) the type and size of wires to be accommodated, c) the type of tooling into which the configuration must be built.

There are coaxial crimpers, ring tongue terminal crimpers, round machine pins and socket crimpers, shield ferrule crimpers, shield splice crimpers, and probably a whole lot more than this writer can think of at one sitting. Each and every one requires detailed training to properly use and assure a quality crimp. Besides crimpers there are pliers; diagonal full-flush pliers, diagonal semi-flush and full-taper diagonal cutting pliers. Add on the long-nose pliers such as round nose, duck bill, long and short and the list keeps getting longer, so much so that it almost becomes overwhelming.

Safety Wire Pliers

Although wire twisters are glorified pliers that can be rotated or twisted by pulling on a spindle clutch, they are a luxury that many mechanics would prefer not to do without. Safety wire can easily puncture fingers—the pliers will help prevent this and also twist smoothly and evenly, throughout the usable length. The gripping area and diagonal cutting edge are not unlike residential wire pliers. The real difference is the spiral twisting tube and locking handles. Slip the safety wire through the hole in the head of the bolt or connector until the wire can be pulled back to meet up with the opposite end. Draw the two ends together until they cross at a 45° angle. The safety wire ends can be gripped, the handles closed, latched, the handles released to lay loosely in your hand, then the push-button clutch is pulled and the wire automatically twists. Grip the handles again, then push the clutch knob back in and repeat the process until the twist is complete.

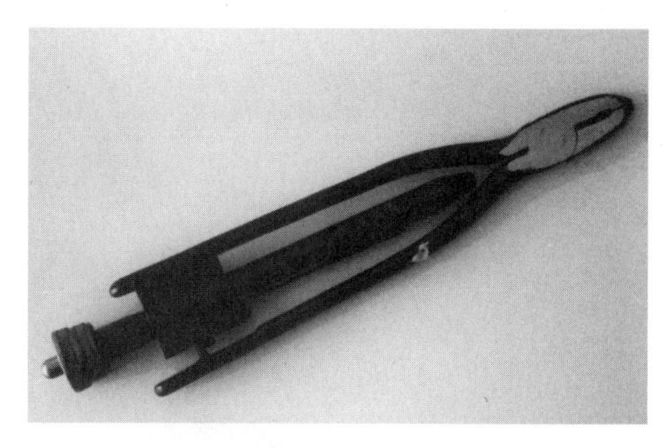

Safety wire pliers; an unique tool for safetying fasteners in an aircraft environment.

There should be six to eight turns per inch; the completed twisted wire should place a tightening effect, opposite to the part's normal loosing direction. In other words, the twisted wire should pull the connector nut or fastener clockwise to tighten. Finish the end by twisting and bending it over to prevent injury to maintenance employees. This tool is amazing; it has been around for a long time and is still one of the best ways to accurately twist safety wire, even in reasonably tight areas. Of course, hand-tightening is the alternative if the area is too restrictive.

Six to Eight
Turns per inch
1
INCH

Hand-twisting of safety wire.

Heat Guns

Shrinking heat-shrink tubing requires a heat gun, but not just any heat gun. It must be capable of handling the abusive conditions offered by aviation. Heat guns made by many manufacturers will work if they are be able to concentrate the heat directly on a section of work, not to exceed .75 inch. This tool will be used to shrink heat shrink tubing onto terminals, splices, or even accelerate adhesive

Example of two-point safety wire and # 47 safety wiring of fasteners on connector shell.

Heat gun for shrinking of heat-shrink tubing, accelerating cure on sealers, etc.

curing. It should be provided with a removable baffle or several nozzle reflectors for concentrating heat on small restricted areas and be able to deliver heat with a temperature range of 750°F+.

A three-way, on/off, cool, and high-heat position switch is recommended; the cool position will prevent thermal shock when the gun is turned off from the high-heat position. Don't lay the gun down hard when hot; the heating element is very soft and vulnerable during this period. Heat guns can be purchased from aviation specialty outlets such as EDMO that carry Weller and GC Electronics. Other manufacturers such as AMP Incorporated or Raychem™ also provide heat guns.

Solderless Crimpers

There is a strong trend towards the solderless terminal for aircraft wiring because of ease of installation and repair. The insulation for the terminal is part of the outer skin of the lug and extends slightly beyond its barrel to partially protect the insulation. There is a metal sleeve under the insulated barrel that grips the insulation of the inserted wire to reduce the stress on the inner conductor. Self-locking ratchet type crimpers such as the AMP Certi-Lok™ part number 169400 with an accompanying die part number 169404 are best suited because of the inability of the tool to release until the crimp is properly completed. If cost is an issue, there is the manual, all-purpose part number 29564 crimper. This tool requires a high degree of skill and common sense. Only if the operator can be consistent from one crimp to another would this be a good choice. Crimping tools like the Ideal model 28-510 and 28-512 handle the entire range of wire gages from 24 to 18 and are less expensive than the AMP Certi-Lok™.

Solderless butt splices are very common, reliable, and popular in aircraft wiring. They give insulation support while securing two or more conductors together in a reliable manner. These spliced connections should always be accessible for easy viewing and repair. To accommodate this, the most commonly used splice is designed to allow inspection of the stripped conductor ends after installation into the barrel. Covering the entire splice is a hard plastic sleeve that insulates the metal barrel from accidentally shorting to other conductors and ground. Between the two crimp sections of the barrel splice is a clear, open space usually called a "window," through which the two ends of the spliced wires should be readily visible. Other insulated splices are the end splice, knife splice, and ferrules.

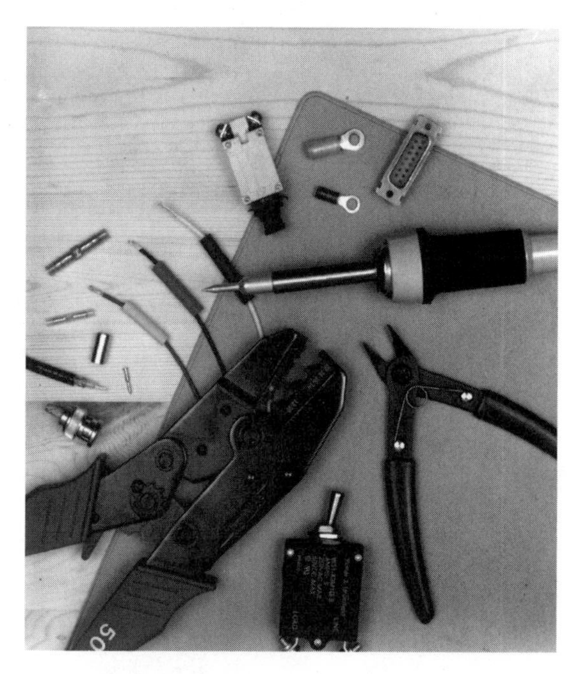

Solderless crimper and assorted tools.

The barrel on most splices is larger than the wire that it is designed to splice and will, when stacked upon each other, cause them to protrude beyond the normal contour of the harness. Staggering the splices will reduce the amount of the bulge and the resultant stress caused by the stiffness of the close proximity of each splice. Because of space constraints, however, it may not be feasible to stagger and some doubling up may be necessary. If it becomes necessary to group several splices in a harness together, clamp the bundle at both limits of the splice section, to reduce the possibility of wire breakage from the stiffer wire bundle.

Regardless of the crimping device, the operator must make sure the tool is fully open by first completing a full test crimp cycle without a terminal. Next, select the wire and terminal; insert into the open jaws, making sure the terminal locator is fully nestled in the tool and properly in position. Place terminal in the respective die (red in red, blue in blue, etc.). Squeeze handles gently until terminal is held snugly in place. Now, insert a properly stripped wire until it butts up against the stop of the tool's locator or if a splice, the internal stop. Start the crimp operation while holding the terminal securely in place, not allowing it to rotate. If the process continues to look good, complete the crimp until the ratchet releases, then remove the finished terminal. During the crimping operation, observe the action from a side angle, if it looks like the terminal is being squeezed away from its proper place in the dies, exert additional hand pressure to hold in place.

INSTALLING SOLDERLESS WIRE TERMINATIONS

1. Determine if the wire selected is correct for the terminal to be used.
2. Strip the wire insulation to the correct length.
3. Insert the terminal tongue-first into the jaws of the crimper until the barrel faces up flush to the built-in tool stop.
4. Insert the prepared wire into the terminal until it you see it extend slightly beyond the barrel opening. Check to see if the insulation is properly inserted into the barrel.
5. Complete the installation by squeezing the handles of the tool until the ratchet releases. Maintain a steady hold on the wire being stripped to insure that it does not slip.
6. Remove the crimped terminal and inspect for correct crimp. Follow by exerting a pull test with index fingers and thumbs of both hands. This is not an official test, but should provide a go/no-go evaluation of what is probably already known, that you have made a good crimp. To detect errors from tool failure, crimping tools should be inspected with gage inspection pull tests on a periodic basis.

BNC Coaxial Crimpers

For crimping BNC coaxial cable connectors, the Super Champ Coaxial Crimping tool from AMP Corporation does an excellent job, requiring three individual crimps to complete the assembly of the connector.

Other tools are available for RG-59 series connectors. The ratchet crimping tool from AMP, part number 607785-1, will provide industry-standard, hex-style crimps on RG-58, RG-59, and RG-62. If the ratchet tool is needed and cost is a factor, consider using the open frame from Daniels or the Certi-Lok™ from AMP, and simply slip in the appropriate dies. This could be the best way to go should you need to crimp a variety of connectors. Custom or dedicated crimpers without removable dies could end up being expensive trash.

Installing quick disconnect BNC-style connectors on RG-58 coaxial cabling for antenna mating, or to allow removal of the King rack without snap-ring removal requires step-by-step preparation. The outer plastic sheath must be removed to a distance of 5/8-inch from the end, leaving the shield exposed. Next, remove 13/64-inch of the shield, preparing the way to remove 3/16-inch of the dielectric. Your final measurements (within ±.03 inches), after stripping, can be made from the cut end of the outer sheath: 5/8-inch overall, 7/32-inch shield to sheath, and 7/16-inch sheath to dielectric end.

After confirming the "cut" measurements, position the thumb and index finger of your right hand on the dielectric, while your left hand is holding the coax close to the prepared area. Rotate the dielectric, in a pivotal action, until the shield has gapped 360° around the dielectric. Before continuing, make sure you haven't disturbed the center conductor. If it was moved, grasp the conductor and gently rotate your fingers, as you pull them toward the cut end. This should assure that the "lay" (strand pattern) returns to normal (note, any maverick strands will increase the overall effective diameter of the conductor, making insertion into the pin or socket difficult).

Whatever method you, the technician, assembler, or installer chooses to remove the dielectric, the end results are what is important. Look for nicked strands; the solid conductor cannot have any nicks, whereas the stranded conductor is more forgiving. However, don't give yourself a license to allow nicks to occur when you have the power to prevent it; sloppiness on your part can escalate until one strand nicked becomes two, then three, and so on.

It's at this point you'll either use ratchet crimpers or the manual type as described earlier to complete the termination. Either way, the center conductor and barrel for the shield must be crimped. If using ratchet crimpers, place the pin into the crimper, bulbous section in the indicated section of the die, insert conductor, and squeeze. Slide the shield barrel down until it bottoms at the coax shoulder; place in crimper until only the "nut" section of the coax connector protrudes. Squeeze and both the shield and knurled portion of the connector will be crimped simultaneously. You're done.

If using manual crimpers for crimping the BNC, the pin or socket follows basically the same procedure, insert pin into indicated crimp area and gently squeeze, this will hold the pin until the center conductor is inserted for final crimp. This is where the two styles of crimpers differ (other than the ratcheting effect); the manual requires three operations to complete the shield and knurled crimp, whereas the ratchet requires only two, as noted in the last paragraph. First, place the tool's crimp die over the knurled section, about 1/16-inch from the end, and squeeze until the pliers bottom out. Then move out to the shield portion, equally spacing the crimp dies from the first crimp and about 1/16-inch from respective end of the barrel.

The above-mentioned connector is called a dual crimp, but single crimps exist (three piece) and are readily available from your dealer; only a single squeeze from the crimper and the pin and shield are both simultaneously crimped, leaving only the installation of the inner portion into the connector body. These are nice, but with "nice" comes the cost. You have to determine if the additional part cost will offset the labor required for assembly.

Stripping of the sheath, shield, and dielectric are considerably easier if using a coaxial cable stripper that can perform all the operations discussed in the preceding paragraphs in basically one simple operation. The only multiple step is the gradual application of pressure through a slide mechanism that applies cutting pressure in increments from three to one, until the outer, middle, and inner sections of the coax are cut. To complete the operation, pull the stripper toward the end of the coax and, "Voila!"—the cable is stripped, each measurement automatically performed without the tedious and hazardous effort of using a razor blade or Exacto™ knife. There are several coaxial strippers available; one fits within the palm of your hand; the other is about the size of a crimper. Some are plastic and others are made from both metal and plastic. The latter is more durable; the cutting blades tend to remain fixed over several strippings.

Soldering creates several problems either relating to heat, or the lack of same. Insufficient heat will cause cold solder joints and high resistance connections. Excessive heat can also cause similar problems when the tin oxidizes from the solder alloy, but the biggest danger is dielectric damage, possibly causing shorting between the shield and conductor or preventing the connector from being assembled because of dielectric swelling. The spring contacts on the quick disconnect, slide-in type, mounted in the back of the rack of some panel-mounted equipment, should be heat-sinked during the soldering process to prevent damage to the spring tension or the insulation.

Preparation of the King quick disconnect connectors (RG124B/U) is pretty simple. Comb out adequate shielding (2.5 inches), split it into two pigtails, and allow them to exit from the cable, on either side of the connector body, just outside the entry sleeve; dielectric removal should be sufficient to just allow the tip of the conductor to drop into the notch provided on the pre-tinned center pin. Pre-tin the conductor and insert into the connector, allowing it to fall directly on the notch, apply your soldering iron to the pin on the side and the conductor should drop in place. Leave the iron's tip on the pin for just a moment, making sure the solder flowed evenly, then immediately remove it (be very cautious not to overheat the center conductor insulator).

Before continuing, pre-tin the connector body, just below the cap, on each side; next, take the excess shield and twist until all the shields are held together, bring them around the coax connector's

The "floating," mating coaxial connector at the back of the radio rack.

body and twist together. Apply your soldering tip to the barrel, just to the side of the shield, and apply solder to the shield until it flows through the shield evenly; the same application should be performed on the opposite side. If you review the illustrations given in the King installation manuals you will note that the shield is terminated on and around the entry barrel; it's my opinion that shield-to-connector-body wrap-around is more secure than this method.

After completing the shield soldering, the shield protruding to the twisted area can be cut off leaving only the shield from the coaxial cable and the point where they join on the connector body. The final step is spot soldering the cap on the body (two places, opposite to each other) and slipping the heat shrink over the entry barrel.

Panel-mounted avionics equipment is usually now provided with a quick disconnect style that semi-floats (slightly loose) to adapt to the variations in rack and radio tolerances. The mating connector on the radio is a simple cavity receptacle, firmly secured to the case, with a white nylon insulator to support the center contact socket and depends on the rack's floating connector. These connectors are designed with a taper to prevent the rack connector from striking the edge of the connector opening. Older methods provide for directly connecting the coaxial connector and cable to the radio; removing the radio required the avionics mechanic, more often than not, to get to the back of the radio to disconnect the cable (this was many times quite difficult and expensive). The new method is as simple as inserting an Allen wrench or screwdriver into the release drive (Allen screw or equivalent) and pulling the radio out; it's almost as simple as removing a light bulb. Some vendors, such as Bendix/King Radio can supply you (for a price) with a ice-tong-style tool to grasp the radio without damaging the knobs or face covers.

How much room/support and degree of bend radius you give the coaxials behind the rack after leaving the connector and clamp are critical issues that must be addressed to prevent premature end-use system failure. The cable must leave in a perpendicular position and route in a gradual radius to prevent structural failure of the dielectric and the resultant, potentially damaging, reflective feedback of the RF energy that flows to and fro between the radio and antenna.

If assembling and installing a panel-mounted avionics package, it might prove prudent to provide a harness support on the radio rack to limit movement after assembly and vastly reduce any potential damage to the either the coaxial cables or the connector wiring. Clamping of the coax should be approached in a cautious manner, the dielectric must be kept intact, but if excess clamping pressure deforms the coax, the distance between the center conductor and the shield will shorten, changing the cable's impedance and its effectiveness within the operating circuit. It is extremely important that precautions be observed when using nylon tie wraps to prevent overtightening; nylon ties can be lethal to the dielectric of the coax. The narrow width of the tie wrap lends itself to a Guillotine effect when over-tightened during installation.

After the cables are installed, finding faults becomes more difficult than if you were to adequately check them on the bench. The only inexpensive way to bench check the cables and the attached connectors is with a simple ohmmeter; this isn't unacceptable if proper assembly procedures were used. If cable runs are short, many of the problems that would drastically affect system operation on larger aircraft with longer runs may not initially show up during normal flight conditions. In fact, they may never show up, but because such failures are possible, greater attention to detail at the bench level is needed to reduce these potential failures. For those difficult-to-find coaxial defects, a reflectometer is necessary. A signal is created by the test set and sent down the cable; if the cable is kinked or shorted, even open, the signal will either be reflected or unimpeded. Either way, an error would be detected.

Should you experience high VSWR readings or a high percentage of reflected power after installation, the probable cause might be:

1. Defective load/antenna—remove and replace.
2. Coaxial cables/connectors shorted or open—substitute, test, remove and replace.
3. Kinked cables—replace.
4. Mislocated antennas—relocate using skin mapping procedure.
5. Poor bonding at antenna base—remove and rebond.
6. Wrong antenna installed—replace with correct antenna.
7. Incorrectly connected coaxial cable to antenna—check connector for damage, install correctly.
8. Too long a transmission line—replace coaxial cable with larger diameter, lower impedance cable.
9. Coaxial cables have incorrect impedance—replace.

Paperwork for Installations and Repairs

You've probably heard this before, but the job isn't done until the paperwork is done. The paperwork attending an avionics installation is considerable. The items you should expect and make sure you end up with after the job is done include: logbooks, FAA Form 337, weight-and-balance and equipment list updates, and an accurate invoice.

Logbook

Our memories may seem to be iron-clad, but in reality, many pilots are inundated with a wide spectrum of problems and responsibilities. Besides, the FAA likes to create a paper trail, and the aircraft logbook does just that. For this reason, pilots should keep track of every part installed. The following list details how this tracking system should work.

1. Keep a complete, detailed, legible logbook entry listing exactly what was installed, the serial numbers of all radios installed, the details on how the work was accomplished, and a reference that an FAA Form 337 was completed and the weight-and-balance and equipment list was updated.
2. After you agree to the installation but before the actual radios are installed, make a note of all serial numbers of the equipment you are buying. The best place to make this note is in your aircraft logbook, next to the page where the shop will log the installation, or in a separate avionics log. This will make it easier to determine if your radio has been stolen and will increase the value of your airplane by showing prospective buyers that you are a stickler for organized, accurate records.
3. This also applies to repairs. Remember that FAR Part 43 requires that any maintenance be recorded in the aircraft records. Lots of people are in the habit of removing a radio, taking it to the avionics shop, then reinstalling it without making any logbook entry. Without a record of what's been done, how will you be able to troubleshoot a recurring problem? Or what if the avionics shop has already replaced the capacitor twice and the transponder still doesn't work? You could rely on your memory, but you should insist on logbook entries.
4. Always bring your logbook with you to the avionics shop and ask for a detailed entry to go along with any repair. You can use a separate avionics log if you want. This will help considerably when you run into recurring intermittent problems, and it can increase the resale value of your airplane. If you don't have your logbook with you when you get something repaired, ask for a typed label that you can place into the log when you get home.

FAA Form 337

An FAA Form 337 must be filled out. At least two copies are required: one for the owner and one to be sent to the local FAA office within forty-eight hours. A Form 337 is required whenever a major alteration is made to an aircraft—any avionics installation should be considered a major alteration. Some repair stations might suggest that they aren't required to fill out a 337; however, FAR part 43 allows repair stations to forego the 337 *only* for major repairs, not for major alterations. Most FAA inspectors concur that any avionics installation, including a mere intercom installation, is a major alteration and calls for a 337 to be completed. The shop might want to keep a copy for its records.

Weight-and-Balance and Equipment List Update

No matter what you've heard, no rule or advisory circular states that items that weigh less than one pound can be ignored for weight-and-balance purposes. If that was true, then how many under-one-pound items could be installed before you'd have to start adding them up and noting their effect on the CG (center of gravity) or empty weight of the airplane?

Every item added needs to be accounted for including lightweight intercoms, wiring, connectors, etc. All items and their weights need to be listed on the equipment list. Shops that take pride in their work will produce a neatly typed weight-and-balance and equipment list update for your records.

Don't accept delivery of your airplane without this paperwork. It really doesn't take that long to accomplish, and it is essential for the legal status of your airplane. The airplane is not legal to fly until the paperwork is completed—it's that simple.

Installation Checklist

Use this checklist as a guide to help you through the avionics installation. You might want to go over this with your avionics shop manager before beginning the process to see if you both agree that these are important points to cover.

The checklist is split into three phases. The first phase is for before any work begins, the second is for about halfway through the job (while the guts of your airplane are fully exposed), and the third is upon completion of the job.

Phase I

1. Meet with the avionics manager, tour the shop, look at examples of work, and contact customers who have had work done by the shop.
2. Finish researching and decide on the equipment you want.
3. Discuss the cost of equipment and installation with the shops you're considering. Get firm written quotes with as much breakdown as possible of the avionics equipment, labor, and miscellaneous parts (such as antennas, wiring, placards, etc.). Discuss what to do if the installers discover unrelated defects while they have the airplane opened up.
4. After choosing a shop to do the work, get a time commitment. Discuss possible penalties for delays in completing work in the allotted time frame.
5. Along with a firm dollar and time quote, ask for a sketch of the planned panel layout, harness routing, and antenna locations.
6. Find out if major wiring harness changes will be needed. Will the shop construct its own harness using MIL-spec high-temperature wire or use the harnesses provided by some avionics manufacturers?
7. What type of clamping of wiring harness will be used, especially over flight controls?
8. How will bare metal be corrosion-proofed?
9. Will the shop do all placarding? What type of placarding? Will it guarantee appearance of placarding?
10. Verify that everything works, or note what doesn't before bringing the airplane in for the installation so you know whether or not something might have been the result of the installation.

11. Get assurances that the shop will provide wiring schematics and assembly notes after the job is done. If the shop is building your wiring harness, a schematic will be needed for anyone to work on the system later on.
12. Will existing avionics and static/altimeter system be tested while the airplane is opened up?
13. Take both close-up and overall photos of the interior and exterior of the airplane and the instrument panel before the jobs begins.

Phase 2

1. Verify that all harnesses are being clamped adequately.
2. Are harnesses clearing control cables and the control column?
3. Have the installers discovered any defects while digging around in your airplane's innards? Will these defects be taken care of? Who will do the work, and how much will it cost?
4. Ask to look at all opened-up areas. Look for anything out of place or obvious defects. Ask questions based on what you have learned from reading this book.
5. Look inside the belly, through the inspection panels that have been opened up in the floor. Is there trash in there that should be removed? Ask if you can do this to save some money and keep your airplane clean. Have all drill shavings, aluminum cuttings, rivets, dropped nuts and screws, and other debris removed. Make sure it is removed prior to reassembly.
6. Confirm that any commitment to use high-quality MIL-spec wiring is being upheld.

Phase 3

1. Ask for documentation of static/altimeter/transponder tests if they were done during the installations.

2. Ask for documentation of testing of all other systems after completion of installation. Better yet, ask if you can be there during testing.
3. Test-fly the airplane with the shop manager, and note any discrepancies that need to be taken care of and that they are the responsibility of the avionics shop.
4. Confirm that all required paperwork has been completed. Has aircraft logbook been signed, with proper dates and tachometer times noted and an accurate description of the work that was done?
5. Is there a listing of new avionics and their serial numbers in aircraft log or radio log?
6. Has the FAA Form 337 been filled out? One copy goes to the owner, and one goes to the FAA (within forty-eight hours). Most shops also keep a copy for their records, although it isn't required by law.
7. Are weight-and-balance data and equipment lists updated and complete? Regardless of what you have heard, there is no "less-than-a-pound" exclusion. If some new equipment has been added, its weight, datum, and effect on the CG must be noted.
8. Take "after" photos and compare with your "before" photos. This reduces the likelihood of differing opinions on prior damage or damage that happened during the installation. It also ensures that nothing is missing or changed compared to the way the airplane was when you brought it in for the work.

It is extremely important that all paperwork be done before the owner/pilot takes delivery. Do not accept an excuse that the shop is too busy and the paperwork will be filled out later. It is a legal fact that the job is not done until the paperwork is completed, regardless of whether or not the paperwork makes the airplane safe to fly.

When you leave the shop, the paperwork goes with you, in exchange for the check that will always lighten your wallet considerably.

Wiring Failure

Wiring fails in the aircraft environment for a variety of reasons: vibration, chafing, carelessness, design flaws, moisture, chemicals, heat, and manufacturing defects. For the most part, the main failure is from the installation process. Even engineering design flaws related to installation can be reduced if proper, good shop practices are followed. It can be said that, "The quality of the product is no better than the design." This is true, but it can be the difference between long-term and short-term degradation of the harness. Although recent revelations regarding commercial aircraft have not publicly spread to general aviation aircraft, the message is still valid. It is time for those responsible for maintenance on aircraft to identify any wire damage to the customer. If located near the fuel system, there are moral and liability reasons to document the deterioration or damage and initiate repairs.

Federal Aviation Administration Chief Jane Garvey said inspections of electrical and other systems of older commercial aircraft are too general and their maintenance is sometimes haphazard. She went on to say that the FAA would implement improvements in how certain airframe systems are inspected, repaired, and documented as aircraft age. According to Garvey, present methods and procedures do not address the problems posed by aging aircraft.

President Clinton's administration released a report on November 15, 2000 that explicitly covered the aging wire problem and suggested some companies that could offer equipment that was capable of locating difficult-to-find wiring defects.

One major concern relating to inspections is the potential for bypassing or ignoring a wiring problem. The number of times failed systems are deferred or signed off as "could not duplicate," also known as NFF (No Failure Found) is almost beyond belief. It is like leaving a bomb ticking, not knowing when it will go off! This is not an exaggeration, it has happened, and recent NTSB (National Transportation Safety Board) investigations are unraveling the gravity of the situation.

Kapton, a wire manufactured by DuPont in the early 1970s, was very popular because it was lightweight and strong. It made its entry into commercial aviation with the Lockheed L-1011 and the McDonnell Douglas DC-10. However, it wasn't long until the U.S. Navy found the Kapton wire was deteriorating. Cracks were found forming in the wire's insulation when subjected to the rigorous conditions of carrier-based flight missions; add the problem of moisture, and the conditions for arcing were apparent.

So what is happening with the wire in commercial aviation? Independent tests and the NTSB have determined that Kapton wire has problems with insulation deterioration over time. This wire is used by many of the heavy iron passenger planes. Tiny cracks, unseen to the naked eye, and chaffing are open doors to moisture, the metal flakes and powder that inhabit aircraft, and other contamination. The metal particles are orphans from the original manufacture and repairs performed over the life of the aircraft. Put these together and the potential for arcing is there. When cracks form in the stiff insulation, it radials around the wire circumference. When the plane cycles through altitudes, moisture forms, arcing occurs, and the moisture is quickly dried by the temperatures from the arcing, leaving a seemingly clean, dry spot. As this process continues, carbon begins to form. Maybe not so it is obvious, but at some point the carbon created by the arcing initiates a potentially dangerous scenario. The carbon from arcing is creating a path or track; the arc then follows the track until it reaches other damaged conductors in the same bundle or ground. The result is flashover, a fire that reaches temperature of over 2000°F. The insulation melts, carbonizes even more, and the process accelerates until gobs of melted insulation is being spewed in all direction from the fire. The now-exposed conductors will sag and can now make contact. The heat damage can be enough to melt the outer skin of the fuselage or any surrounding metal. If fuel vapor is present, it is all over!

The insulation used in wiring can be toxic when exposed to high temperatures like the aforementioned. It should be mentioned that Kapton wire is no longer being installed. However, for a long time, it was used extensively in military and civilian aircraft, including NASA vehicles. A very large percentage of the planes flying today have Kapton wire strung throughout their innards, including the president's plane, Air Force 1. In fact, by this release, Air Force 1 will have been revamped with new wiring. Navy aircraft have already faced this problem head-on and have rewired many of their planes. Of course, Navy planes are subject to harsher environments than most other aircraft, and the need may be greater.

The number of hours or cycles on a plane is used for reference purposes by both the FAA and commercial aviation. For all practicality, however, the deterioration is related to environmental flight conditions over time. This is further compounded by the human element. For general aviation aircraft, all the annual inspections, repairs, and scheduled replacements add to the movement of the harnesses but do nothing for preventing the problem. This is because no regimented requirements specifically look for the elements of wire failure noted here by the FAA on general aviation aircraft, and understandably so.

Depending upon the current being carried and the type of insulation, an explosive sound created by rapid heat may reach levels similar to that of Fourth of July firecrackers or higher. The crew should be constantly aware of any change from the norm. Of course, smoke is the most commonly known indicator, but sudden crackling in the headsets or speaker, loss of power, flight deviation, or suddenly erratic instrument readings may be a sign of electrical damage. If there are any fuel vapors or other hazardous materials in the vicinity, they might very well explode. If near the fuel tanks, it may be too late!

Educating the customer, slowly but firmly, is the first step to creating an environment of mutual understanding. Critical aircraft systems should be removed and replaced with new wiring. Just adding insulation is not enough. The miniature cracks that occur in Kapton wire are almost invisible to the eye, but not to moisture and the conductive contaminants that can bridge the gap between a safe flight and disaster.

Pilots, crew, and passengers skirt the edges of potential disaster every time the plane taxis out of its parking space or hanger. Just like the automobile, the plane has tangible risk factors that are understood by aviation professionals, but not necessarily by "civilians."

Any time there is a potential for failure, additional measures must be implemented. Call it paranoia, call it heightened awareness, call it anything you like, but an ounce of prevention will prove to be a pound of safety. Heat creates undue stress on the materials that form the protective covering of the wire. Combine the chemicals found on aircraft, along with the broad spectrum of heat and cold, and the tendency to fail escalates dramatically. Of course, squeeze the wire under a clamp, bracket, or similar condition, and the insulation will eventually spread out, away from the conductor until it shorts to ground, or to an adjoining wire. Putting too many wires in a conduit promotes failure because of the pressure that builds between the wire insulation and the conduit inner walls. Engineering should design for the conduit's internal diameter versus the diameter of the proposed bundle and allow for only a 40 percent fill. This would be based on circular mil area. What happens here is added equipment and therefore more wiring is added until the limits are exceeded.

Vibration and design are factors that the FAA is finally finding to be an issue in extensive wire failure. General aviation aircraft have a lower potential for large numbers of human life lost caused by any kind of accident. This is simply because fewer passengers make up the numbers. This is not to say the problem isn't there. It is the responsibility of each technician and mechanic to inspect a lot closer and make sure the client is fully aware of the potential for failure. Actually, it should be put into the aircraft's logbook if considered a safety-of-flight item.

Manufacturing failure could mean that the wire was incorrectly manufactured, improperly labeled, or just not tested to ensure that it does not have dielectric failure and will meet impedance specifications—another less discussed failure. Checking a spool of wire is called periodicity (repeated flaws); it checks for reparative structural return loss (SRL) from things such as a defective pulley, wire extruder, and the like. Although most manufacturers only test an occasional roll of cable, it is unlikely that defect will turn up in those not checked. This assumes that the equipment manufacturing the wire is functioning properly. As a receiving facility, the wire should be source inspected and a SRL graph provided. A SRL graph will not only tell you about periodicity but also related changes in capacitance, impedance, conductor size, and cable dimensions. Because of periodicity, even minor flaws will be magnified over distance.

When observing system failures in signal and digital circuits that evade detailed troubleshooting techniques, the problem might be traced to those repeated flaws. It is possible to simply draw your fingers along the wire and "feel" the variations in insulation thickness. Substituted wire is one sure bet for failure if specifications are not the same as required by the design. If wiring is routed near or in environmental conditions that are detrimental to the insulation, failure is sure to occur. Oil, fuel, and other chemicals will eventually damage many wire materials, even if the material is predominantly Teflon.

One potentially damaging, but preventable factor that is difficult to call out on drawings is chaffing. Essentially, harnesses, plumbing, equipment, and other hardware should not come into hard contact with any structure or each other. The question remains: How do you define hard contact? Actually, at no time should equipment or racks come in direct contact to any part of the plane, except where that contact is clamped. At least 1/4-inch or more is recommended for adequate clearance for fuselage skins or adjoining structure.

Shock-mounted equipment must take the movement of the radio during vibration and erratic motion of the aircraft. The same is true for clearance of plumbing, especially those carrying oil, fuel, and oxygen. There should be no physical contact between adjoining plumbing, wiring, or structure. Wires are insulated, but the insulation isn't impervious to damage from sharp edges, heat, or excessive pressure, such as that imposed when clamped with nylon ties against a metal surface. Never use plumbing for primary support. Spacing hardware methods of separation are acceptable, but at no time should the plumbing carry the weight of the harness.

Although the wire insulation may resist heat, it may not be that unyielding to chafing or surface abuse from contact with brackets, radio chassis, etc. Make sure that wiring is properly clamped from beginning to end. Don't overtighten harness mounting clamps. Although it may seem that everything is working right, the pressure exerted by the clamp may eventually cause the conductor to work through the insulation and may potentially reveal itself to other conductors, causing a wire-to-wire short. Where the cable or wiring exits the connector, a clamping method must be used that provides adequate support to the cabling, however, it has the negative effect of inserting a "stress riser" that concentrates considerable force to a small area. In single- and multiconductor-bundled conductors, this isn't a major problem, but it may potentially present a serious problem with coaxial cables where the dielectric is compromised.

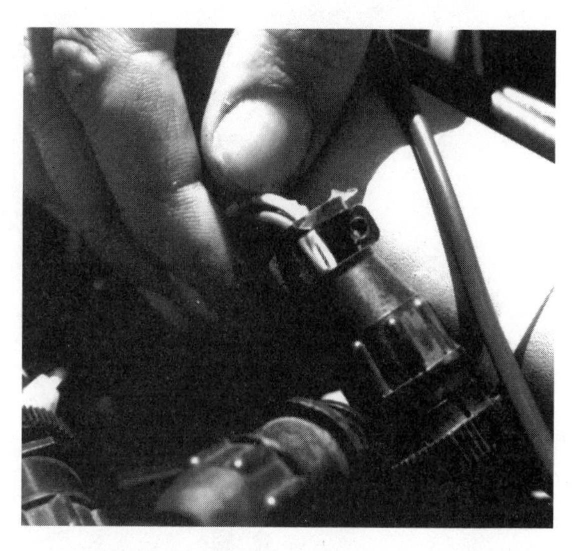

Clamp was overtightened until it squeezed the conductors through the insulation, causing two wires to short together.

Harnesses must be tied no more than 15 inches apart in long runs, except for breakouts where the minimum spacing should be 3 inches. Keep in mind that these are minimums; other factors such as harness shape, lack of adequate structural supports, and location demand more ties located closer together. Anytime the harness support is over 12 inches (should not be more than 24 inches), there should be closer tie groups. Usually, the closest spacing would be about 1 inch and that is needed on the wiring behind the instrument panel and at breakouts. Breakouts should use a "Y"-style tie to pull and keep all three harness tie points together.

Clove hitches and square knots are the industry standard for creating harness ties. This method requires that nylon string (called tape by the vendor) be used for the knots. First, lay one end of the string over the harness, bringing it down and back under the harness (on

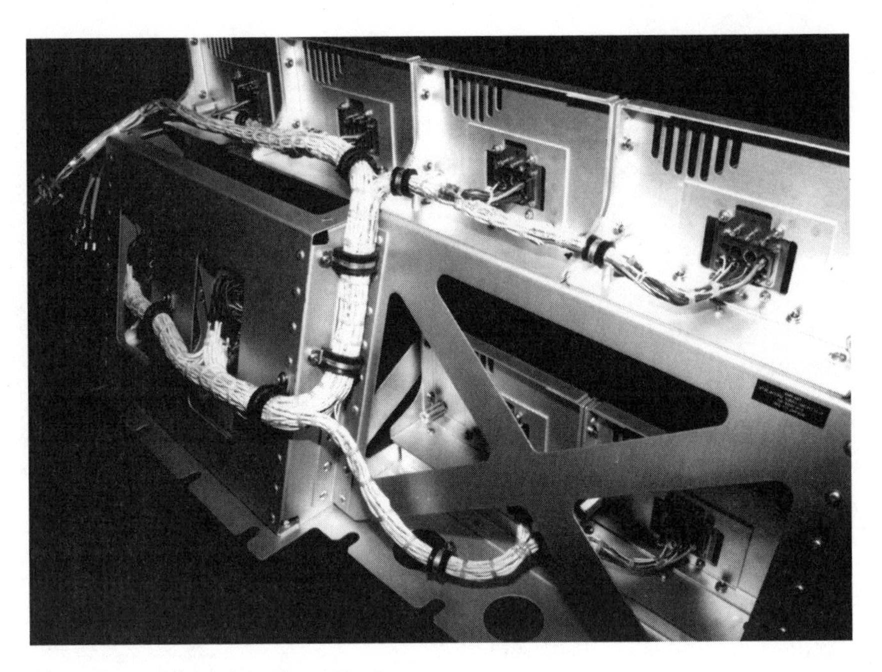

Video equipment rack harness routing and breakouts.

the right of the incoming string) toward the installer. Continue up toward the installer and back over, dropping it across the incoming string and down around until it can be inserted under the first incoming crossing string. Finally, a square knot is tied over the clove creating a secure knot. A second knot is needed to prevent the first from working loose. After the tie is complete, cut excess off to about 1/2 inch from knot.

Yes, nylon ties or straps can also be used for general harness assembly and installation, but they must be limited to areas outside of con-

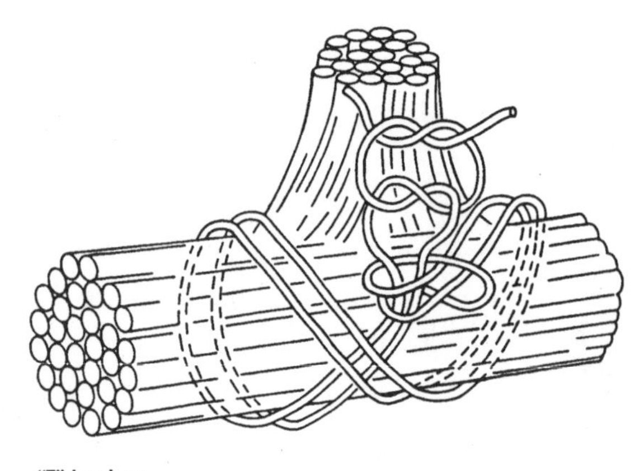

"T" breakout.

trol cables and associated hardware. This is especially true behind the instrument panel. Do not forget, nylon ties are also strong enough to exert a damaging, choking effect around the harness. Initially, there won't be any problems, but over a period of time, the wire strand conductors will draw closer and closer together until shorting takes place. Nylon straps are prone to breakage and the harness could fall into moving parts causing an in-air accident. Additionally, nylon ties should not be used to support harness in the overhead cabin. Although it may not cause an accident, it could cause a system failure should the sudden drop yank wires loose from bulkhead connectors.

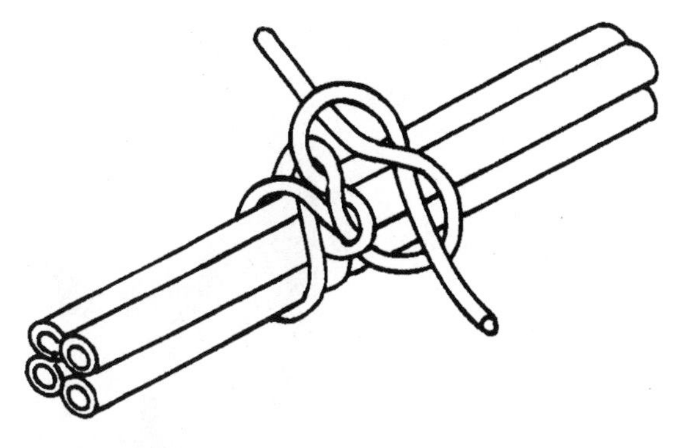

Clove hitch tie-in.

Only metal clamps with rubber insulation should be used for support harnesses over control surface cables, pulleys, and the like. These are commonly referred to by the manufacturer's name, Adel. When clamping a harness with an Adel clamp, assure that the clamp section is below the fastener holes with the mounting point being up.

For all harnesses and coaxial cables, the potential of chaffing is very real. Preventive measures must be put into place to prevent such an obvious danger. This means adequate clamping, edge protection, such as grommets, and nylon "caterpillar" devices should be installed where the harness passes thought lightening holes, across shelving, and even other harnesses. In all cases, the intention is to prevent relative abrasion. This means that there must be some motion; even small relative motion between the harness and abrasive surfaces can effect damage to the insulation of a given wire or coaxial cable. There are documented cases where nylon straps on engine harnesses wore enough of a scar into the engine strut that it had to be changed. Don't

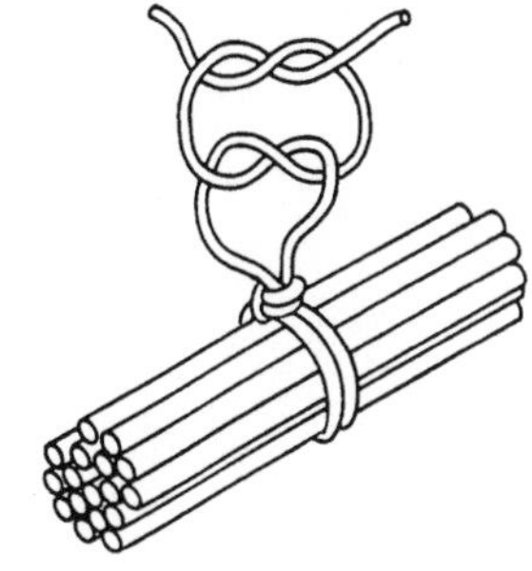

Clove hitch ties.

Adel clamp used to support harness; however, note that the clamp was installed upside down and may eventually break if there is adequate stress.

underestimate the patience of hidden errors. They have all the time in the world—you don't.

The majority of coaxial cable used is usually RG-58/U and works quite well between the radio and the antenna system. If putting in a new radio(s), the old antennas should be replaced with those suggested by the vendor in the installation manual. If not, a mismatch may result that will vastly affect transmission and reception range. Should this be a retrofit, the older the wiring, the higher the probability of existing damage from nicks, chaffing, or cracking. For this reason, new wiring must used for a new installation. Additionally, replace the existing circuit breakers; they also age and could present a liability to the operating life of the new installation.

How much room/support and degree of bend radius you give the coaxials behind the rack after leaving the connector and clamp are critical issues that must be addressed to prevent premature end-use system failure. The cable should exit the rack in a perpendicular (90 degrees to the connector) position. It must route in a gradual radius to prevent structural failure of the dielectric, not to mention the potentially damaging, reflective feedback of the RF energy that flows back and forth between the radio and antenna.

If assembling and installing a panel-mounted avionics package, it is necessary to provide harness support on the radio rack to limit movement after assembly. This will vastly reduce any potential damage to the either the coaxial cables or the connector wiring. Clamping of the coax should be approached in a cautious manner. The dielectric must be kept intact, but if excess clamping pressure deforms the coax, the distance between the center conductor and the shield will shorten, changing the cable's impedance and its effectiveness within the operating circuit. Especially important precautions should be observed when using nylon tie wraps; they can be lethal to the dielectric of the coax. The narrow width of the tie wrap lends itself to a Guillotine effect, cutting through, or depressing the area within the cable or wire.

After the cables are installed, finding defects becomes more difficult than if you were to adequately check them on the bench. The only inexpensive method for bench checking the cables and the attached connectors is with a simple ohmmeter. This is acceptable if proper assembly procedures were used. If cable runs are short, many of the problems that would drastically affect system operation on larger aircraft with longer runs may not initially show up during normal flight conditions. In fact they may never show up (as far as we know), but because such failures are possible, greater attention to detail at the bench level is needed to reduce these potential failure.

Corrosion and Contamination: The Unseen Enemies?

Most avionics systems are engineered to render reliable operation under difficult weather and temperature conditions. However, several factors are uncontrollable by the black box manufacturer's design that will create failures, either intermittent or permanent. This may be poor solder connections, improper crimps, ineffective bonding, insufficient cooling, ignored corrosion, or something as simple as a pinched wire. The technician must be experienced and knowledgeable to adequately maintain the avionics in a modern aircraft. This means learning as much as is humanly possible before soldering one wire, crimping one terminal, or bolting two parts together. It comes down to one insurmountable fact: The aircraft's survival depends on those responsible for its maintenance.

Many times those things that are taken for granted end up being a major obstacle to proper system operation. An example might be the direct replacement of existing outdated radios, such as the KX-170 or TKM's MX-300V, with a fully digitized, modern-style radio that will satisfy FAA frequency requirements. The old radios might still be working but simply need to be replaced, however when the new TKMs are installed, the unthinkable happens! The radio fails to work. The new owner of the TKM is now frustrated and cursing his or her decision to do a direct change-out and not totally replace the avionics. "I knew I should have bought a new installation," the owner complains, teeth grinding as he or she continues to build a four-letter vocabulary.

What happened here? As convoluted as it may seem, the answer is very simple. The mating contacts have formed into a permanently squashed or depressed position; as long as the radio wasn't removed it most likely would have worked for a long time. Although the contacts were previously mated with that radio, a new radio would have a different curve to the shape of the contact. Additionally, there is always the problem with corrosion and tar buildup from smoking. What happens is the tobacco contamination creeps in and round the exposed contacts. When the radio is removed and a new or second radio is installed into the same rack, the contacts may not mate on the same exact contact points. A little elbow grease (work), and this could be a happy occasion. By the way, do not move the attached harness; it is stiff, and movement could cause difficulty in finding failed connection.

Panel-mounted radios are prone to problems not found with other aircraft avionics systems. One particular source for failure can be traced to tobacco smoking. The tars and other contaminates produced by smoking are known to coat microphone jacks, radio connector contacts, internal radio switching, and even sealed indicators. Yes, it took time to create the damage, but the harm is quite evident and widespread and may take some effort to correct. *Caution:* Before cleaning, make sure the power to the avionics is shut off. When cleaning radio contacts, it might be necessary to individually lift the contacts to reshape to the original configuration.

A strong disclaimer: Please do not attempt this operation unless you are prepared to replace the connector. You must first clean the contacts and, of course, have the proper tools.

Besides, a good cleaning may be the only maintenance needed. Use approved contact cleaner that won't affect plastics and or collect dirt after cleaning. A good electronic contact cleaner that does not leave a residue should do the job.

Should it be necessary to "lift" the depressed contact, considerable caution should be observed. The plastic lip that keeps the contact from slipping from the connector can be easily broken if too much vertical pull or lift is exerted. The contact can then slip off and not engage the radio. If viewing the contacts through the empty open rack with a flashlight, it should be easy to determine if the contacts have become depressed, therefore lacking the necessary spring action to maintain contact with the mating connector contacts. Lifting the contact should take place at the highest point on the connector contact. Necessary tools are two dental cleaning probes; both with 1/4-inch, 90-degree tips or ends. To perform this procedure, firmly press one probe against the end of the contact where it rolls over the connector contact support ridge. This is to keep the contact from slipping from the retention lip and thus chipping the plastic.

Next, slip the other probe under the contact where the greatest gap is seen. Lift very gently and move the probe back and forth slightly, creating a rolling action; it won't change the height much, but it doesn't have to. Do this for each contact. The back and forth motion will tend to stretch the metal slightly and keep the new contact curve in place. Once this step is complete, install the radio and call it a day. Again, be extremely careful not to over-lift the contact, the retaining plastic lip will break!

Metal corrosion is the deterioration of metal through a chemical process that effectively damages or weakens the affected material to an unusable level. It may not be obvious; the corrosive deterioration may be unseen, lurking beneath the surface of the protective plating or paint. This problem is generally more invasive in coastal areas. Water, especially when mixed with salt, accelerates the process. Other corrosive agents are industrial chemicals that are a natural enemy of metals. Actually, the only metal (other than aluminum) in an aircraft not easily attacked by corrosive elements is gold. That is why it is used for mating contacts on printed circuit boards and wiring connectors.

What does corrosion look like? It varies with the material. It may initially be seen as a puffy, blistering presentation along rivets and next to antennas. The most common metal in an aircraft is aluminum. Pitting, accompanied with a white or grayish powder is the tell-tale sign of corrosive elements attacking aluminum, magnesium structure, avionics racks, and radios. However, don't be confused by the white oxide that forms on aluminum alloys. It will usually measure between .001- to .0025-inch thick and is not harmful. Actually, this creates a protective coating that prevents further corrosion. Copper and brass, a copper alloy, when corroded will have an antique green coating. The green coating may be desirable when choosing Spanish or Southwestern lamps, but it is totally unwelcome in aircraft. Other metals such as steel will display a reddish deposit called rust. If the steel is cadmium-plated, the corrosion can possibly penetrate the plate and be traveling under the protective layer. This may occur unnoticed until eventually revealed by the tell-tale signs of paint or plating blistering. Next, cracks may manifest themselves in the nooks and crannies of the pitted areas with the part eventually failing.

Many terminals or crimp-style connectors such as splices and lug style have an open-end design at the insulation support and where the wire strands exit. This is where a green corrosion may be the most evident, especially along coastal areas. This corrosion is difficult to remove because it has most likely crawled along the side of the strands and sequestered itself deep under the insulation. Of course, sealed connectors should be used, but that wasn't the case on many aircraft. Corrosion removal and sealant may be the most economical approach to correction. Where the corrosion has penetrated too deep to remove, the wires will have to be cut and new sealant terminals installed. Using the right chemicals, this may not be too difficult.

Switches and fuse holders are yet another target for corrosion. Switch connections may have lost their corrosion proofing because of handling; the same is true for the fuse holder. Repeated insertion and removal of the fuse scratches the clips and removes the plating. Periodic inspections should uncover these types of problems.

There are two corrosive activities that can have disastrous effects on aircraft: One, is a direct chemical attack with corrosive liquids or gases such as battery acid. This is less of a problem than the electrochemical process that occurs in high moisture and salt conditions. Preventive measures can alleviate the direct chemical attack, but the latter is more prevalent and therefore presents a greater threat. In this case, the observant eyes of the technician may encounter a potential failure before it becomes a problem.

To prevent corrosion from forming between steel and aluminum components, cadmium washers should be installed between the two metals. In addition, the fastener used to bolt the metals together should be cadmium or primer-coated. Where two dissimilar metals are mated together, the potential for galvanic corrosion is highly probable, especially if moisture is present. This presents a very serious condition because of the concealed conditions. Destructive damage can unfold and not be discovered until too late!

This is why it is critical to reprocess any surface that is disturbed during maintenance or equipment replacement. Removal and reinstallation of an antenna disrupts a sealed environment under the antenna base. Previous installers may have left the installation in less than desirable shape, therefore the reason for antenna removal.

Inspect the following areas for corrosion and poor bonding:

1. The black box (radio) to the equipment shelves
2. Metal-bonded shelves to structure
3. Junction boxes to shelves/ground structure
4. Engine mounts
5. Inverter mounting
6. Strobe light power supplies
7. Shock mounts
8. Antenna base to airframe
9. Flaps, elevators, rudder gear doors, between hinged components and structure
10. Cowling
11. Plumbing
12. Static wicks
13. Open terminals and mating contacts
14. Switches
15. Riveted junctions
16. Circuit boards (internal electronics)

A lot of corrosive action can take place on aluminum without incurring serious damage. However, it is still damage and must be stopped and corrected. The process is relatively slow and provides a wide window for discovery during inspections. One sign of corrosion may be noise in the receivers gradually increasing over a span of several months—particularly at hinge points and poorly sealed antennas. As mentioned earlier, salts will accelerate the corrosive process, but simple preventive measures will prevent further damage. Alodine 1200™ (chromic acid), primers, paint, and other surface coatings go a long way to create a corrosion-proof barrier. This barrier between the outside environment and the metals that form the structure and supporting appliances of the modern aircraft is always vulnerable to outside damage. Applying these protective coatings requires training and experience.

Circuit boards or expensive switch systems that are corroded beyond minor cleaning should receive a complete cleaning. ChemTech*, a company specializing in corrosion removal, can effectively recover electronic components at a fraction of the cost of replacement. The cost for most circuit boards might be around $75. The larger and more complex the boards are, the higher the cost. When compared to a great number of circuit boards costing thousands of dollars, $75 is cheap. The process uses cleaning machines and heated cleaners (CT-1, CT-2, and CT-3) to soak and spray components until they are almost like new! A corrosion-proofing coating (CT-4) is then applied to protect against moisture entry and potential corrosion. For general cleaning, CT-1 will work quite well. This would work on surface oils from handling or the environment. However, unless using heated tanks with CT-1, excessive contamination and corrosion removal is not as efficient. ChemTech will be glad to help solve avionics corrosion problems. *Note:* Their process also protects the recovered component from future corrosion. In many cases, the boards will function as designed without repair. The only damage might be an erosion effect from the corrosion. On a side note, if it is impractical to get the corroded components to a recovery facility within twenty-four hours, they should be submerged in distilled water. Make sure any installed batteries are removed and properly discarded.

Most modern aircraft are painted a basic white acrylic or polyurethane with stripes and serial number added as the plane is assigned to a customer. If the aircraft is unpainted, then it is most likely an alloy with an Alclad® (aluminum clad) lamination. Relatively pure

*ChemTech's phone number is 703-360-8004 or Fax 703-360-4468. They are located in Alexandria, VA (1800 Diagonal Road, Suite 600, Alexandria, VA 22314). Other phone numbers are: 703-549-1001, 800-268-6189, or Fax 703-549-1003. They can also be contacted by e-mail at chemtech@idsonline.com.

aluminum is extremely resistant to corrosion but lacks the strength to handle the stresses necessary for aircraft equipment racks and other structural components. For this reason, other metals are added to form an alloy. Where the alloy is not to be anodized or painted, a thin lamination of pure aluminum is added during the manufacturing process to provide a layer of protection. Clad aluminum provides the best of both worlds—strength *and* corrosion resistance. However, should the thin protective coating be mechanically removed or penetrated by sanding or scratching, the corrosion-sensitive alloy can be potentially invaded by caustic elements. Equipment racks are usually anodized to keep maintenance cost down. Close inspection should reveal the low-gloss, textured effect created by anodizing. The problem with anodizing may be poor bonding.

The area where the antenna is mounted is a target for corrosion, especially on older planes. This is because moisture finds refuge between the surface of the antenna and the aircraft skin. A good inspector should find this problem quickly. If the corrosion has advanced to the point of severely pitting the skin, a doubler will have to be added to the exterior of the plane. If done properly, the doubler will be invisible to the casual observer. The antenna would then be reinstalled per specifications.

The avionics metal bond honeycomb shelf is a good example of an applied RF bonding technique. Caution is critical during the bonding process. The bonding procedure can actually create the conditions for corrosion. Each step of the bonding is relatively simple, but it should be noted that those points of weakness are perfect for corrosion. A honeycomb shelf is composed of an upper metal skin, a layer of adhesive, metal expanded core (honeycomb), another layer of adhesive, and finally another metal skin. Because the skins and core are separated by an adhesive that is nonconductive, it won't be possible to electrically bond both skins without providing a method to connect both together, then to airframe ground. Bonding is accomplished by wrapping a strap from the top skin around to the lower skin in at least two physical locations, on opposite ends of the shelf.

Now that the two skins are effectively RF-bonded to each other, you can move to the next step and run two more straps, on opposite ends of the shelf, to the airframe structure.

Aluminum Preparation for RF Bonding Straps

1. Remove protective coverings such as paint, primers, or corrosion with non-reactive non-metallic abrasives or wire bonding brush (*Corrosion alert!*).
2. Follow up with MEK or suitable low-residue solvent to remove residue.
3. Apply Alodine 1200™ anti-corrosion coating; work in with an acid brush or nylon abrasive pad such as Scotchbrite™.
4. Remove after three minutes (after golden color appears) with a damp sponge or cloth by blotting or spray. Do not rub; this will remove the soft coating.

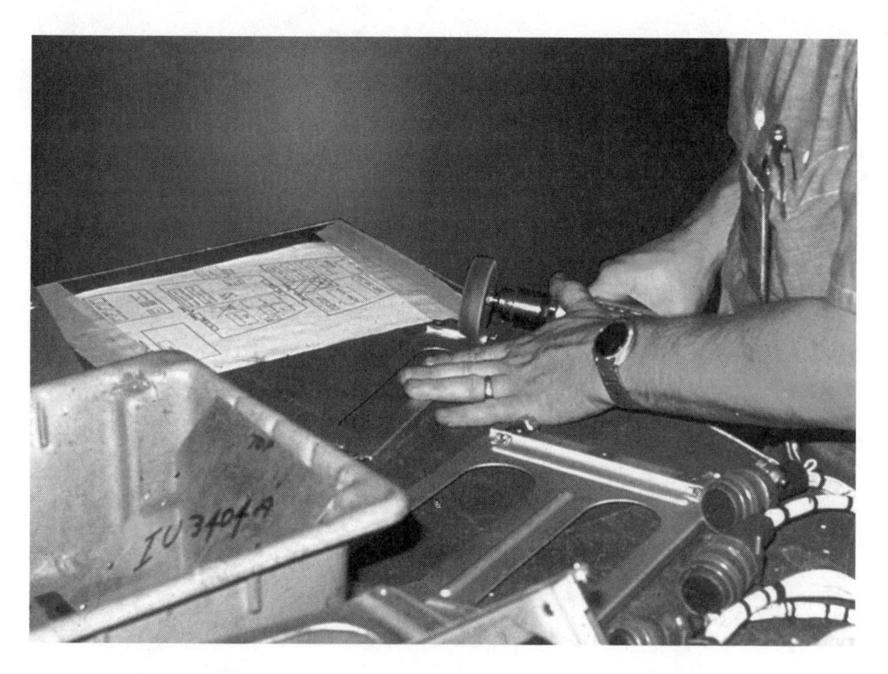

Equipment racks and other component racks being prepared for corrosion proofing. After Alodine is applied, the equipment racks will be installed to the shelf.

5. Do not remove prematurely and don't leave on any longer than a half-hour. Failure to remove within one hour, as is recommended by Ray-Chem, will result in the Alodine 1200™ surface becoming brittle (*Corrosion alert!*). If you do allow the Alodine 1200™ to harden, reapply another coating of Alodine 1200™. If the time limit exceeds two or three hours, it is highly recommended that the old anti-corrosion coating be removed followed with a new application of Alodine™. In most cases, it is highly recommended that 20 minutes is an optimum time frame for applying the solution followed by the avionics equipment. Once 24 hours have passed, the solution becomes nonconductive.
6. Install the "U"-shaped bond strap over the top and bottom of the shelf.

Radar shelf bonding to airframe through two (1-inch x 5-inch) bonding straps. Strap is shaped like an "U" and installed to bond the top and bottom of the shelf skins together and to the airframe.

7. Install the applicable fasteners through the hole in the upper strap through the shelf and finally through the lower end of the strap. Thread on cadmium-plated washers and nuts and tighten to the appropriate torque. The core will need a strengthening agent such as epoxy to prevent crushing of the shelf.

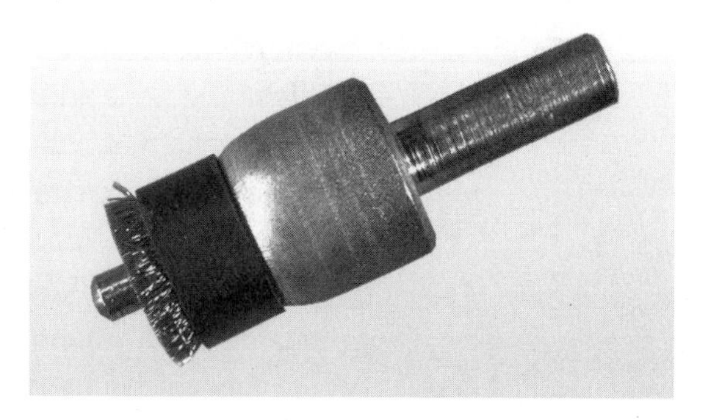

Bonding brush.

8. After the installation is complete, apply a finish coat of the original protective coating or clear lacquer to prevent corrosion from entering the exposed area (*Corrosion alert!*).

A similar corrosion-proofing procedure is used on magnesium and steel, but with different chemicals. With magnesium, brush the area with chrome pickle solution and immediately remove with water. Follow with mounting the brackets or components. With steel,

Example of a completed bonding application with an installed insert.

remove any contaminant films such as grease or oil with MEK or Trichloroethene and allow to dry. Do not remove or damage the cadmium plate that will serve as the non-corrosive surface that will mate to the respective components.

Drill the hole for the strap, insert the bonding brush pilot into the hole, activate drill motor, and apply pressure, removing surface coating, leaving a smooth polished metal surface.

Apply corrosion-protection chemicals; install terminal or strap and insert screw, washer, split-lock washer, and nut. Tighten and apply original surface coating or clear lacquer.

Antenna Install Specifications

Because the avionics radio depends so heavily on the antenna and its low impedance to the aircraft structure, the same bonding procedure described above must be also applied to the base of the antenna element. Not only will a poor impedance path cause inefficient antenna operation, so will the capacitive effect caused by gaskets, corrosion, or any other non-conductive dielectric films. Removal of gaskets and paint is mandatory. They cause more problems than the good for which they're intended. Only if contours or irregular-surfaced antenna bases are encountered would a spacer-type gasket be required and then only if bonding washers are provided through the gasket thickness to the airframe. There is a new bonding material that solves some of the gasket problems and the RF bonding. To find out more, please contact the author of this book.*

If an antenna currently has a gasket, remove and bond directly to the airframe, *except where the manufacturer has introduced a conductive gasket, such as the Stormscope® or Strike Finder®*. Remove the surface paint or corrosion to an area not to exceed 1/16-inch outside of the flange of the antenna base. Cleaning the antenna base may also be required. Reinstall antenna after corrosion-proofing and check with a bonding meter to measure less than .0025 ohms. Some of the newer antenna systems have conductive gaskets supplied; confirm before using.

**Ed Maher:* emaherr@worldshare.net or (760) 322-3614

After the antenna is installed, apply a 1/16-inch bead of manufacturer's recommended sealer around the base, holding the sealer gun at a 70° angle to the mating point of the antenna and airframe. Move the gun away from the starting point, the gun angle down, and the tip moving away from the installer. This prevents air bubbles from forming under the bead. Finally, the installer should use a moistened finger or spatula to form a 45° fillet between the base and the skin.

Recently sealed COMM antenna. Note small fillet of fresh sealer around the antenna base.

After all bonding procedures are completed, a bonding check should be conducted using a central ground point in the fuselage to connect one end of the tester. On some aircraft, a common ground location inside the fuselage is used, preferably on the wing carry-thru. This point carries out electrically to the wings and fore and aft of the fuselage. Look for readings of less than 1 ohm. Actually, a resistance of .0025 at all points is recommended, to insure a minimum of problems.

Follow the detailed systematic list for reliable antenna bonding procedure and corrosion control:

1. Determine size of the area to be bonded after the corrosion repair process. Outline the area on the fuselage or wing to be bonded. Stay 1/16-inch within the antenna's base footprint. Remove paint or primer and treat the bare aluminum with Alodine 1200™ (chromic acid) or Dow 19™* anti-corrosion solution (for magnesium). Next, prepare the antenna, being careful to bond as much of the base as possible.
2. If the antenna is supplied with a gasket, remove and discard.
3. Install doublers and nutplates.

*Dow 19™ is a registered trademark of Dow Corporation.

4. Using an air pressure-operated sealer tool or a caulking-style gun, apply sealer around the cutout for the connector. After mounting the antenna, apply a bead around the outer periphery edge of the base.
5. Make sure the antenna is installed within 20 minutes to the prepared surface and secured with fasteners. It is important to tighten the fasteners to their respective torque values.
6. Clean off any sealer that has squeezed out, being cautious not to leave any voids or gaps at the base of the antenna, and if necessary, add additional sealer.

Corrosion control requires the expert intervention of a knowledgeable avionics technician or mechanic. This section reflects only a sampling of the total requirements for bonding and equipment mounting.

Treatment of Corrosion-Affected Areas on Alclad® Aluminum

POLISHED ALUMINUM

1. Remove surface skin contaminates such as dirt or oil by using mild emulsion cleaners.
   - Water-emulsion cleaners (general purpose)—to be used on both painted and unpainted aircraft.
   - Solvent-emulsion cleaners—heavy grease and oils; some can soften paints. There are two versions; one is nonphenolic and the other a phenolic base. The nonphenolic is safe on rubber and paint, but the other is not.
2. Use metal polish to clean the surface, but be very careful to use only polish intended for Alclad® surfaces.
3. If there is resistance to cleaning, a stronger cleaner for aluminum is available that will hasten the process and recover the normal luster to the aluminum.

4. Before using any of these compounds, initiate a test on an area that can be corrosion-protected by another process if this should fail.
5. Treat the surface corrosion with an inhibitor-imbedded solution. If this chemical is not commercially available in a given area, use a solution of sodium dichromate and chromium trioxide. Leave on affected area for about fifteen minutes, then remove using mineral-free water and wiping dry. Complete by applying Alodine 1200™ to the footprint of the antenna.
6. The antenna or radio tray is then installed per the procedure outlined following these corrosion treatment steps.
7. Finish by applying aircraft-applicable waxes around the antenna or radio tray. This will protect the area around the repair.

With painted surfaces, the removal process can be more aggressive. Paint thinners or strippers can be used to remove residual paint, then Alodine 1200™ (chromic acid), primers, and paint will be applied to complete the job once the corrosion is under control.

PAINTED ALUMINUM

1. Clean the fuselage surface with a suitable aircraft cleaning solutions.
2. Use drycleaning solvent to remove grease and oils.
3. Apply water-soluble strippers to remove paint.
4. *Caution:* Before using any of these compounds, make sure the process is performed in an open-air environment or a paint-stripping booth with a positive airflow.
5. Make sure all rubber areas are protected from the solvents, especially tires and seals. Many solvents may degrade rubber, especially synthetics.
6. Steamclean and police up the mess.

It is highly unlikely that an avionics technician or mechanic will ever perform the extent of corrosion control described above, except possibly in the case of major modifications where large sections of the fuselage will require extensive cutting. For the most part, corrosion control will be isolated to equipment racks and antennas.

Antenna arrays represent less than 5 percent of the entire avionics system, but are 100 percent the reason the radios are able to work. Whether trying to pick up (talk/listen) the tower during approach, performing boondocks navigation, or simply flying in bad weather, it can be the difference between life and death.

Check out the following basics; corrosion can play a major role in the operational capability of a communication system. When the microphone key is keyed, a sinusoidal alternating voltage is produced and transmitted at the selected frequency. The amplitude of the created voltage determines the output power of the transmitter. When this voltage reaches the antenna, an electromagnetic field is emitted from the antenna. Upon reaching a receiving antenna, a resultant, duplicate, but smaller voltage is produced. This signal is now amplified and processed through the receiver. Factors that can affect the signal accomplishing its trek to the antenna or from the antenna, or the reciprocal, is any mismatching of impedance between transmission lines, antennas, and the transmitter/receiver.

Obviously, impedance-matching of interconnecting cables is paramount to an efficient operating system. The SWR (standing wave ratio) of an antenna, such as the communication radio's, should be as close to 1.0 as possible. This can be tested with a Bird wattmeter and a 50-ohm resistive load placed in-line of the transmitted signal. A SWR reading of 3:1 means that 75 percent of the signal is being transmitted, but 25 percent is being reflected; this is an indicator of potential failure. The antenna must present itself as a proper load to the transmitter system so that a minimum loss of transmitted power is achieved. The active resistance of the antenna or the coaxial cable that is attached to it cannot be measured with an ohmmeter, as the resistance specs of the antenna are configured at the respective frequency of the transmitter. The coaxial cable can, however, be continuity-checked with an ohmmeter or test light for point-to-point continuity. Corrosion under the antenna base will drastically affect the SWR and the shape of the transmission pattern.

The following seven factors must be considered when developing and installing a highly tuned antenna system:

1. Installation characteristics
2. Mechanical durability
3. RF efficiency
4. Proper placement
5. Precipitation/lightning protection
6. Corrosion prevention
7. Proper bonding

Shortcuts that can potentially damage the antenna should be avoided. Should the gasket supplied not accommodate the minor fuselage contour deviations then cracking or delamination of the

Corrosion at base of antenna. The paint is blistered by underlying corrosion.

base of the antenna may occur. Although your aircraft may not carry commercial passengers, the antennas should, for your benefit and safety, meet or exceed the standards of TSO requirements. These are standards set up by the FAA for any equipment that is to be used on commercial passenger aircraft. These standards insure that the antenna will be built to the highest possible standards in all of the seven categories listed above.

When the military determined that over a third of their maintenance activities were related to wiring and related systems, it wasn't a surprise when quality assurance programs went under scrutiny. It was also no wonder that such attention was given to wiring quality when you consider that a little over a foot of wire figures into a pound of gross take-off weight. The attention that the military gave to wiring did not spill over to the civilian community; this is why so many failures are attributed to wiring. Add in the cancer of corrosion, and you can expect more wiring failure. Corrosion is slow but inevitable. As mentioned before, it can enter the space between the individual conductor strands of a given wire. The same is true for terminals. If the terminals that form the link between the equipment and the wiring are not crimped with the correct tools, breakage or high-resistance connections may occur. Adverse environmental conditions and the wrong type of connector opens the door for corrosion.

Commercial aircraft still enjoy the infamous reputation of high maintenance because of wiring-related failures. These failures start at the factory level with "human" errors that some may consider to be a "normal" and "expected" part of doing business and continue into the real world of transporting people and products. Factories spend a considerable amount of time and money in an attempt to reduce these in-process errors, however, the system is not foolproof. Applying these programs to field would go a long way to further reduce the possibility of errors reaching the end customer. Competition has eroded what the FBOs are able to do. It simply costs a lot to set up training sessions and many shop owners have decided to get by on the skills they can hire. Again, there is the problem. Availability of technicians and installers remains extremely low, with starting wages at an all-time high. The plane owners should find those shops that have training programs as a normal part of doing business. Yes, the cost may be higher, but how do you judge the cost of life? Planes seem to be falling out of the sky in the months this book was written. Pilot error, possibly, but keep in mind that the same mentality exists in the ranks of many shops.

This doesn't mean to say that failures won't show up or that increased awareness will solve all future failures. The aircraft is a complex animal, but with an increased layer of quality awareness, the degree of failures and downtime should diminish to acceptable levels. But what is an acceptable level? This is a reasonable goal determined by administration that provides the maximum effect with minimal cost. Before any unrealistic goals are achieved, acceptable goals *must* be established. Improve the communication between those who build and those who pay for the results, and improvement can always be expected.

Preventive and corrective measures can, and must be, taken to solve potential and present problems before they develop into a serious and dangerous life-threatening situation. What are these measures and how can we go about implementing them? One of the first steps is inspection. Regular detailed inspections must determine if corrosion exists and then appropriate action to correct it must occur. These inspections should be scheduled with a check list being used to assure each and every item is evaluated. Also, determine if any system malfunctions or inaccuracies exist and trace the cause. Should that cause be human error or design deficiencies then the method of correction should be clear to the trained technician.

CHAPTER 5

Microphones, Headsets, Intercoms, and Speakers

Microphones, headsets, intercoms, and speakers are the basic devices that allow pilots to interface with their communications radios. At the same time, they can be a significant source of the problems the technician and aircraft owner will encounter.

The Microphone

A microphone must deliver the human voice clearly during reception and transmission. Because the microphone must survive environmental conditions such as heat, cold, high humidity, dryness, vibration, and rough handling, it must be rugged and dependable. If buying a new microphone, ask around to see what other pilots are using. While low-cost and low-weight might appeal to you, find those pilots who have microphones that are over a year old and are still operating reliably. Only well-designed, sturdily built units will sustain the rough environment of being thrown around an airplane trip after trip.

Microphone by Electrovox™.

There are three basic types of aircraft microphones in use; the carbon, dynamic, and electret. Each type can be designed with noise-canceling capability; however, the noise canceling is more compatible with dynamic and electret microphone.

Carbon Microphone

The carbon microphone has been around for a long time and is the simplest and least expensive of the three types. You might have used one like the familiar black Telex 66C.

In a carbon microphone, a diaphragm is attached to a small chamber containing densely packed carbon granules. As the sound waves from your voice strike the diaphragm, the carbon granules are compressed. The compressed carbon particles alter an electrical current that is being applied across the carbon chamber. The changes in the electrical current are amplified within the COMM radio, then transmitted.

Drawbacks of the carbon microphone include inherent noise, only fair frequency response, plus as they age, the carbon granules become more tightly packed together they stick, causing unwanted signals and reduced capacity. You may have seen pilots whacking their carbon microphone on the dashboard or throttle console, which is a valid technique because it loosens up the granules, making the microphone operational again, extending its life for a little longer. *Hint:* scrape together a few bucks and buy a new set; they are getting old.

If the carbon microphone is not the noise-canceling type, ambient noise can enter the voice grill along with voice sound waves and drown out or overwhelm transmissions, making them unreadable.

The primary advantages of the carbon microphone is low cost, durability, and resistance to radio frequency interference (RFI) and electromagnetic interference (EMI). You can virtually throw this microphone around the cockpit without damaging it; in fact, that might even help prolong its useful life. For the pilot, this is a pretty good choice for an inexpensive emergency backup microphone.

Dynamic Microphone

The dynamic microphone is composed of a wire-wound coil connected to a diaphragm. As the speaker's voice strikes the diaphragm, the coil moves in and out along a core magnet, producing a voltage that is amplified then transmitted. Dynamic microphones are heavier than electrets, but they are reliable, virtually immune to RFI and have excellent frequency response.

Electret Microphone

For noise-canceling microphones, the electret is the best choice. It is small, light, and produces the best voice transmission. Electret microphones are also highly resistant to outside RFI and EMI, plus they are inexpensive and highly reliable.

Electret microphones feature a fixed backplate and a charged mobile plate that moves proportionally to the movement of the diaphragm. The charged/fixed backplate design of the electret microphone promotes clear voice transmission. The noise-canceling feature works via an additional port in the microphone through which ambient noise can enter. Ambient noise enters through the additional port and hits one side of the diaphragm while simultaneously entering through the voice grille and impacting the other side of the diaphragm. Both sources of noise cancel each other out by hitting the front and backside of the diaphragm equally. This leaves the voice striking only one side of the diaphragm, generating an electrical signal as the distance between the fixed and charged plate varies.

Microphone Problems

Microphone problems are frequently the cause of many transmission complaints, so it's important to learn how to isolate the microphone as the problem before spending too much time and money on other components.

Most avionics shops have a microphone test box for testing keying and modulation. With the microphone plugged into the test box, the tech can easily check for intermittent wiring problems by wiggling the cord where it enters the microphone and watching the test meter for intermittent operation. If the test needle is slamming over into the red zone on the test meter, the microphone is overmodulating and, if possible, needs to be adjusted. The meter will also show if the microphone's output is too low for proper transmitter operation. Past experience has found that pilots have purchased microphones that simply will not work with older avionics systems. Don't overlook this as a possibility. This is easy to check; just use one of the tried and true electrets.

SOME POSSIBLE MICROPHONE PROBLEMS

1. *Gain adjustment set too high.* This occurs when excessive amounts of cockpit noise are being transmitted. Readjust the gain on the microphone and confirm proper reception with someone listening to your transmission. You might run into this problem when installing a new microphone that is differ-

ent from the old one. If your airplane already has a dynamic, electronically amplified microphone, replace it with the same kind of microphone. In any case, make sure you check the output quality of a new microphone whenever you replace an old one.

2. *Distorted signals.* This occurs when the signal develops instability before being transmitted. Causes include speaking too loudly, misadjusted gain, poor quality microphone, damaged microphone element, or even poorly installed antennas, defective shielding, and poor bonds. The instability introduces other frequencies in the voice signal and changes the character and clarity of the signal.

3. *Your microphone technique may be causing the problem.* Placing the microphone near the comer of your mouth rather than directly in front of it and close to your lips and speaking at a normal level will reduce the chance of poor transmission. Try an experiment. Using a remote microphone and a tape recorder, speak a fifteen- to twenty-word sentence into the microphone, holding the microphone directly in front of your mouth. Try the same sentence again, but this time hold the microphone at the corner of your mouth. Now play back the tape and compare the sound on the two sentences. The first sentence, where you held the microphone in front of your mouth, will demonstrate the injection of sharp "S," "P," and "T" sounds. The second sentence should be relatively free of such sounds and be distortion free, an important requirement for aircraft transmissions.

DEFECTIVE PUSH-TO-TALK SWITCH

It's easy to tell if a push-to-talk switch simply isn't working. You won't hear the familiar "click" as you depress the microphone button or the push-to-talk switch. It is much harder to detect a short-circuited push-to-talk switch because such a failure can render your radio inoperative on any frequency. This is also known as a "stuck mic." It can also prevent other aircraft and transmitters within range to be shut off from communicating, so it's important to learn how to detect this problem.

If the plane has two microphone jacks (the receptacles into which you plug the microphone) and they are wired in parallel to each other—quite common in most installations—a shorted microphone button on the copilot's side can activate the key line and continuously transmit a carrier signal. This will infuriate everyone trying to use whatever frequency you're transmitting on and prevent you from using your radio.

If you suspect this problem, the easiest way to troubleshoot is to switch microphones. With a good microphone on each side, you should be able to transmit and receive normally. This is another good reason to carry a spare microphone. If either the pilot's or the copilot's push-to-talk switch is stuck, then you have a real problem and might have to shut your COMM radios off to prevent blocking everyone's transmissions. This is also a good reason to carry a portable transceiver, for emergencies.

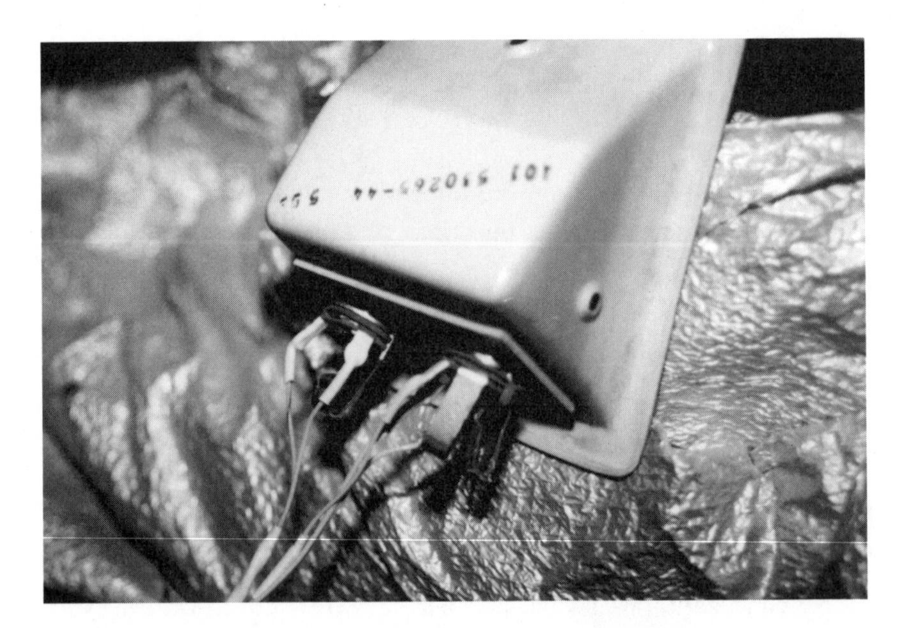

Recessed microphone and phone jacks removed from pilot's side panel for inspection.

JACK PROBLEMS

A common microphone installation problem is installers using the incorrect microphone jacks. The jack must have the correct detent distance to match the microphone plug; otherwise the plug won't make the proper contact and may fall out. To check for this, simply grasp the body of the microphone plug and tug gently. If the plug pops out easily, the detent on the tip of the jack is probably not mating adequately with the spring-loaded contact finger on the jack. Even if installers use the correct jacks, the contact finger can become bent or misaligned during subsequent maintenance or normal use.

Although it's better to replace an incorrect, misaligned jack, you can correct a mismatch between the jack's contact finger and the plug by reforming the contact finger on the jack. Remove the plug from the jack, and make sure the jack's diameter is compatible with the microphone plug. Bend the contact finger down slightly, then back up, then finally back to the horizontal position, which should bring the finger back into alignment with the detent in the tip of the microphone plug. If the misalignment isn't excessive, putting a slight curve in the long contact finger that mates with the plug detent will often do the job. This should be done by a qualified technician, not the pilot.

Intercom jacks should always be installed with a floating ground, which means that the jack itself is insulated from the airframe and not installed so the metal jack can contact the metal on which it's mounted. Use insulating washers to isolate the jack, unless the mounting structure is nonconductive. Then terminate the jack's ground at the intercom. This is the accepted way to reduce potential sources of audio noise.

WIRING PROBLEMS

Testing and repairing your microphone can lead to disastrous results if a few simple preventive steps aren't followed. If a meter is to be used, make sure the active element is not in series with the test leads.

Before tearing open a microphone, check the wiring for intermittents by plugging in and operating the microphone while wiggling the wire where it enters the microphone with the mic plugged into a test box. A good indication that the wiring may be defective is signs of cracked insulation around the microphone cord, especially where it enters the microphone.

After carefully removing the screws from the rear cover, gently lift the microphone body away from the front section, exposing the push-to-talk switch and the dynamic element. You'll see where the wires enter the body of the microphone and the push-to-talk (M) switch contacts. M-switch contacts can become contaminated with nicotine if smokers fly in your airplane, plus corrosion can build up on the contacts through normal use. Clean the contacts with some electrical contact cleaner, available at a local electronics store or at a well-stocked pilot shop.

With a volt-ohmmeter, test for continuity between various areas to confirm that the wiring is in good shape. There are five places to check on this microphone:

1. With one probe on the barrel of the mic jack (the portion nearest the wire) and one probe on the tip, depress the push-to-talk (PTT) switch. The meter (VOM) should indicate continuity, confirming that the M switch is operating correctly and that the key line should work.

2. Place one probe on the center section of the jack (ring) and the other where the wire enters the element (from the switch). Again, depress the PTT switch and check for continuity; this confirms one of the audio wires. These first two tests are the only ones where you'll check for continuity while depressing the PTT switch. On the remaining tests, you'll just be checking for continuity between the wires themselves.

3. You can't test for continuity through the dynamic element, so check from where the wire enters the dynamic element to the ring on the jack to confirm the other audio wire.

4. Check continuity from the tip of the jack to the solder joint at the switch.

5. Check continuity from the barrel to the solder joint where the cable is clamped to the mic body.

The shielding on the cable terminates at the jack and should be isolated from the jack. If you suspect shielding problems, you can check for continuity between the shield to any part of the jack. There should be no continuity on this test.

If you find a defective wire and feel reasonably confident about repairing it, remove the cable and cut it back beyond the damaged area. This is assuming that it is long enough to repair as opposed to replace or splice. Remove the outer plastic sheath and determine the length of each wire needed, then cut the wire and strip the insulation from the wire. Pre-tin the wires and solder to their respective points. Be careful not to use excessive heat, which can easily damage electronic circuitry and plastic parts. Recheck the soldered joints with the VOM and reassemble the microphone. Plug it in and test it by asking the tower or unicom for a radio check.

Some newer microphones like Electro-Voice's have a removable cord. A small plug-in connector at the base of the mic body makes changing a defective cord a much simpler job.

Before Taking Your Microphone to a Shop

Any time you have one of the following problems, try a backup microphone first before opening your wallet.

1. COMM receiver doesn't work. (Transmitter might be keyed.)

2. COMM will not transmit. (Microphone key button might be defective.)

3. Tower can hear carrier signal, but no voice. (Could be broken wire at the jack or microphone.)

4. Transmissions intermittent.(Could be a defective key botton, an intermittent break in the mic cord, or a bad solder joint.)

5. Transmissions weak, garbled, breaking up, or noisy. (Possible defective microphone, intermittent tip line, or damaged shielding.)
6. Hum in speaker when mic keyed. (Floated microphone jacks with insulated washers may be shorted to the airframe.)

Buying a Replacement Microphone

There are a wide variety of microphones to choose from. Some are designed for right- or left-handed people, for people with low- or high-pitched voices, and for helicopter versus airplane use. I recommend the most expensive aviation microphone you can afford, designed for the type of aircraft you'll be flying most. You'll find the more expensive microphones deliver cleaner audio and have a long-lasting reliability.

Look for microphones that are TSO'd; this assures you the microphone meets minimum quality and durability standards. You'll want to make sure your microphone has excellent noise-canceling capability, is amplified, adjustable, comfortable to use, and resists external interference. Warranty support is also important. Some manufacturers such as David Clark are notable for this; they support their products for nearly the life of the product, while others aren't so helpful. Ask other pilots, aircraft owners, and avionics shops before making a purchase decision.

Headsets

Headsets have come a long way from the hard plastic earpieces that used to fit into the classic barnstorming leather helmet. They are available in a wide range of prices and capabilities, from simple, earplug-type, lightweight designs to the expensive active noise reduction (ANR) headsets pioneered by Bose and now offered by a number of manufacturers at even more reasonable prices.

There are two basic types of headsets to consider. One is the lightweight design that consists of portable headphone-type earpieces that have good sound quality but don't reduce noise at all. The other is the full-blown noise-canceling headset with the large earmuff-type headphones; these offer considerable noise-reduction capability. If you fly airplanes with extremely quiet cockpits, you might find the lightweight headset more comfortable for long trips.

The noise-canceling earmuff headsets do a great job of reducing harmful noise, but they do so by clamping your head in what feels like a vise after a long flight. Be sure to try on a pair before you buy to see if they are comfortable for more than a few minutes. You'll find that headsets with liquid-filled cushions are more comfortable and seal better to your head than headsets with foam-filled cushions. Also, the liquid-filled cushions mold better to your head if you are wearing glasses. Headsets usually have a noise-reduction rating that is useful for comparing their effectiveness, but nothing beats a trial flight in a noisy airplane to make comparisons.

Model H10-30 David Clark headset.

Active noise reduction (ANR) headsets are becoming more popular, especially now that more manufacturers are entering the business. Most work by generating an "antinoise" exactly opposite in phase to the noise coming from the engine, wind noise, etc. The actual noise and the antinoise, if properly generated, cancel each other out, leaving a remarkably quiet environment for your ears. ATC and intercom communications come through loud and clear with ANR headsets, and they are definitely worth the investment if you fly a lot in noisy environments.

ANR headsets can be used as portable headsets if you fly a variety of airplanes. If you own an airplane and are buying ANR headsets, have your avionics shop hard-wire the power source and control module into your panel so you won't have wire "spaghetti" all over the cockpit. The portable systems use a cigarette lighter plug for power. One benefit of these systems is that if the noise-canceling function stops working, the headset will operate like a normal headset, still protecting your ears from damage, although not quite as comfortably as when the unit was working as designed.

As humans we can hear or detect sounds in a range from 16 Hz to 20,000 Hz. Not as good as the animal kingdom, but it does the job. To communicate we speak in a range of 250 Hz to 2,000 Hz. Why the broad difference? Simply put, hearing is defensive, and as human animals we need to hear danger before it finds us. That is why it so important to have the right equipment to protect us from excessive noise and still hear the sound of danger.

As one who has lost much of his hearing to the high-pitched intensity of U.S. Air Force KC-135 jet engines up close and personal, it isn't too hard to make a serious request. You must take the time and money to get the right ear gear for working around and flying aircraft. Aircraft noise provides a broad spectrum of danger signals such as the stall-warning horn, squeal from mechanical parts, or the uneven rumble of an engine with fuel or ignition problems. Having two ears allows us to determine distance and direction. Losing either ear or compromising their ability to do the job vastly reduces our ability to function safely in the aircraft environment.

Inside your ears are cells that resemble hairs; if vibrations strike the cells and vibrate them at too high a level, they are damaged, and even destroyed. It would be like running through a room full of chandeliers with a baseball bat, swinging as hard as you can. Flying deaf is not ideal, so if you enjoy your hearing, don't take chances with it.

The Occupational Safety and Health Administration (OSHA) has established 90 decibels (dB) as the maximum sound level humans can tolerate without risk of injury. A commercial jet's engine can produce sound in excess of 130 dB. Within just a few seconds, these levels can easily cause irreversible hearing loss. Cockpit sound levels vary from 80 to 100 decibels, depending on engine size and how well the plane is insulated.

For working in and around heavy-decibel areas, ear plugs are recommended as simple, effective, and inexpensive protective devices. Although different manufacturers make units with different comfort levels, almost every one will reduce the potentially damaging sounds by 15 to 20 dB, while still allowing you to adequately hear spoken words.

Although a little heavier, the noise-attenuating headsets seem to do a better job by covering the bone-conducting areas around the ears. With noise-attenuating headsets, tests show a reduction of 20 to 25 dB, however, for maintenance personnel, this is a pretty expensive noise protection device. For headsets to reduce the noise effectively, you must have a good seal between the ear cups and your head. Manufacturers usually use two methods. One is a foam-filled cup and the other is gel-filled; however, if you must wear glasses, the gel-filled is a better choice because they tend to fill the gaps around the ear pieces more effectively.

Although more expensive, the ANR headset provides an excellent seal around the ear to prevent noise from entering. The electronics monitor the ambient noise and produce an out of phase counter wave that cancels the noise. Prices range from about $80 for a mod kit to over $600 for top of the line ANRs. One company manufactures an ANR for less than $200, which is a very good price.

Aircraft headsets are designed for the proper 600-ohm impedance rating for aircraft electronics. Regular stereo headphones aren't designed for the same impedance and using them could cause damage to the avionics. Sporty's Pilot Shop sells a plug-in impedance-matching device that adapts 8-ohm stereo headphones to the 600-ohm avionics (see Sources at end of book for address and phone number).

While most aircraft headsets can be used in any aircraft, some are designed specifically for the high-noise environment in helicopters. Don't waste the extra money for a helicopter headset unless you are flying helicopters.

Aircraft headsets are manufactured using most microphone types, but the electret has proven to be the lightest and most reliable for headset use, especially in terms of noise-canceling capability.

Headsets with dynamic microphones are cheaper but might not give as satisfactory a result. Again, if you can, try before you buy, and ask around for advice from fellow pilots.

The same problems that plague microphones cause trouble with headsets. Be careful not to coil your headset wires up too tightly when you're done flying; just let the wires flop loosely in your headset bag to keep them in good shape. Don't leave the headset on an instrument panel on a sunny day. In fact, keep your headset away from the instrument panel altogether because the magnets in the speakers can harm your aircraft's compass.

Many headsets offer easily exchangeable parts, so if you're planning to keep your headset, you may want to buy one that is easily repairable.

Intercoms

Nothing will improve the comfort of your airplane when flying with passengers more than a good intercom. Without an intercom, noise from the engine and prop and slipstream will overtax your ears and drown out your voice, causing you and your passengers to yell at each other to be heard. Flight instruction without an intercom is extremely difficult, and it is surprising that most manufacturers do not automatically install intercoms in their training aircraft. Intercoms can make the training experience much more enjoyable and productive, not to mention prevent hearing loss among future pilots.

Cockpit noise levels are usually 80 to 100 dB, and the average person in airplane environment speaks at 95 dB. If there isn't at least a 10 percent difference between the voice decibel level and the ambient noise level, the voice will be smothered by the noise. So you can see that in some of the noisier aircraft, you must really raise your voice to be heard. It is more than likely that a lot of people have quit flying lessons because the noise and the yelling made the experience so uncomfortable.

An intercom is the least expensive and easiest avionics device to install. There are a variety of brands available, and you can purchase a good-quality, portable unit for under $200.

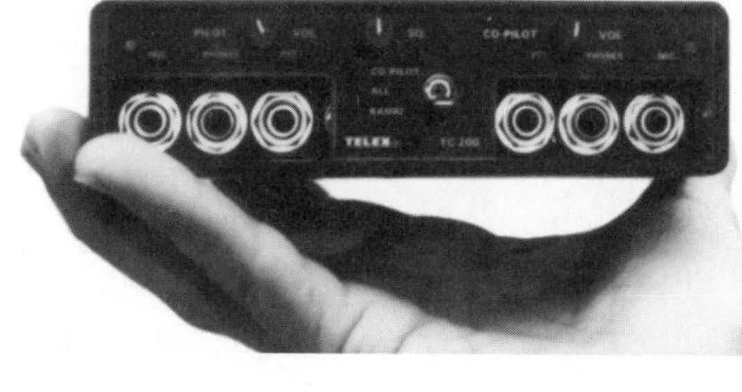

Telex intercom.

Some Features to Look For in an Intercom

VOICE ACTIVATION

Most modern intercoms have this feature. When you speak into your microphone, only then does the intercom pick up your voice and transmit it to whoever is connected with you. Outside noise isn't transmitted. If you have a chance to test the intercom before buying, make sure the first few syllables are not cut off before voice activation kicks in.

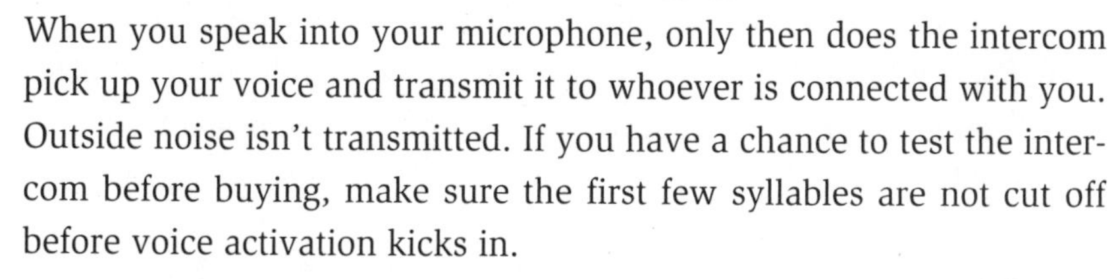

NUMBER OF STATIONS

You can buy intercoms with two, four, or six stations, depending on how many seats are in your airplane and how many seats you want to wire for intercabin communications.

PERMANENT OR TEMPORARY INSTALLATION

If you are, say, a flight instructor who teaches in a variety of non-intercom-equipped aircraft, you might want to buy a pair of headsets and a portable intercom. The same goes if you rent airplanes that don't have intercoms. The drawback to a portable intercom, however, is the proliferation of wires in the cockpit. It's not pretty, and you have to be careful not to tangle the wires with something important, like the flight or engine controls. A permanently installed intercom stays with the airplane and cuts down on the number of wires snaking around the cockpit. One disadvantage to the permanent

David Clark DC Com 50.

installation is the extra cost of installation. Much of this is caused by having to remove lots of upholstery to route new wires.

SEPARATE VOLUME AND SQUELCH CONTROLS FOR BOTH PILOT AND COPILOT

People have their own comfort levels and will enjoy greater comfort if they can set volume and squelch to their own satisfaction.

HORIZONTAL OR VERTICAL PANEL INSTALLATION OPTION

This allows flexibility in deciding where to squeeze the intercom into your instrument panel. If you can't install it both ways, you might find that the unit you've chosen won't fit in the remaining real estate.

SEPARATE PILOT/COPILOT AND PASSENGER CIRCUITS

This allows you to shut off the passenger circuit so you and the copilot can communicate without having to listen to the grumbles and groans from your backseat flyers. This can be very useful when you're shooting an approach to minimums.

Intercom Installation Tips

For mounting a panel-mounted intercom, locate a point on the instrument panel that you can comfortably reach and where there is enough room behind the panel to fit the intercom and some wiring. Usually, you'll find an open spot on the copilot's side of the panel works best. You won't have to actuate the intercom controls that often, so it doesn't need to be right in front of you. If there is absolutely no space at all, you can have your installer build a little bracket to mount the intercom just under the bottom of the panel, as long as it doesn't interfere with your legs and full movement of the controls.

Intercoms usually come with stick-on placards that adhere to the panel after the intercom is installed. Ask your installer if they offer an alternative method of placarding the intercom, like silk screening the panel or making up a laminated plastic tag (see placarding discussion in Chapter 4). You want to try to match the intercom installation with the rest of the panel so it looks natural and blends in with your existing avionics.

Ask your installer to use MIL-spec wiring. It is more tolerant to soldering-iron heat and has better quality protective insulation. Some intercom manufacturers supply their own color-coded wiring harnesses with their products. Don't use them until you and your tech have inspected the wire and the quality of the solder joints and determine that the quality is better than what the installer can build.

If the intercom manufacturer recommends that mic and phone jacks be lifted above ground (not grounded directly to the airframe through the mounting surface or instrument panel), consult your installer to make sure that's how the jacks are installed. This requirement prevents noise from entering the audio system via the grounded panel, especially in the case where the bonding of the panel to the airframe becomes loose, deteriorates, or is accidentally removed during maintenance.

Intercom wiring should not be tied or mounted to any heavy current-carrying cables like the large cables going to the battery or to pulse equipment coaxial cables (transponder and DME coax). Routing intercom wires close to these cables has historically been a major source of noise.

Whenever possible, resist the temptation to route the intercom bundle on or next to the main electrical harness running or any other current-carrying cables. This increases the potential of inducing interference into the intercom wiring. If there is no choice, they should be attached in such a way that separates or spaces the intercom wiring from the current-carrying cables.

To ensure backup operation in case of intercom failure, the audio amplifier should be wired directly to a set of microphone and headphone jacks labeled "Emergency Mic/Phone." This will allow normal operation of the radios using the spare hand microphone you'll always be carrying. All audio wires should be twisted, with shielded pairs for optimum noise suppression, and the shielding terminating at the source radio.

Intercom Jack Installation

When consulting with your installer about where to place the rear-seat headphone and microphone jacks, don't just opt for the most convenient spot. Many installers, for example, like to remove the ashtrays from the sidewall in modern Piper singles and twins and install a little plate in place of the ashtrays. The same is true for the Beech Duke (Model 60) and others.

More important, of course, is whether the installer's recommended jack placement will work. Take the ashtray example. Have you ever sat in the back seat of a Piper wearing a headset plugged into jacks where the ashtrays used to be? I've found this is a big mistake many owners and their installers make. The back-seat passengers might find themselves pressed against the jacks, making them uncomfortable, especially if there are two people in the back seat. Additionally, this can damage the headphone plugs and ashtray jacks.

One solution for Pipers is to install a doubler in the overhead air-vent tunnel that runs along the ceiling. The jacks would be installed in that tunnel, making them easy to reach and out of everyone's way. If that is too many headphone wires dangling in the air for your comfort, another option is to install the jacks on the aft baggage compartment bulkhead. That way the wires can rest along the back of the seat, with excess wire stuffed into the baggage compartment.

Yet another possibility is using 90-degree mic plugs or recessing the jacks into the ashtrays, at 45-degree angles, similar to the Beechcraft Duke (Model 60) and many other planes. Each of these methods will work, but some are better than others.

Remember that it's up to the aircraft owner to make the decision, but it should complement the shop's recommendations. Although the pilot is the one who is going to have to live with the installation for a long time, some installations are simply not practical. Some installers initially opt for the easy solution, like replacing the ashtrays, so it's up to you to make your ergonomic tests and make your desires known. One reason the installer will take the path of least resistance is they are trying to keep costs down. This is one key reason that the aircraft owner will need to become part of the decision process.

After the installation, both the shop and the pilot should perform a final test of the intercom installation by powering up the avionics. Next, turn on the ADF and tune in a station's audio or a radio station plus ATIS on one of the COMM radios. When the microphone is keyed, the audio from both sources should mute. Intercom audio levels do need to be calibrated to the aircraft and microphone you're using, so make sure the installer does this as a normal part of the installation.

Speakers

Speaker operation is relatively simple. A small wire coil inside a magnet vibrates whenever a voltage from an amplifier is applied. The coil is attached to a paper cone, and when the coil vibrates, it makes the cone vibrate. The vibrating cone creates sound waves that the pilot hears as a voice from the speaker.

Speaker Problems

Frequently pilots don't notice their speaker has failed because they are long-time users of headsets and never use the speaker. It's a good idea to check the speaker occasionally to make sure it's working because it is as much a backup for your headset, as the headphones are a backup for the speaker.

If the speaker is suspect, plug in your headset and listen for clear and undistorted audio. If the audio is low from either radio, you'll have to have the radio checked, but if it is low when listening to the speaker, and not when using the headset, then the speaker is the problem.

Speaker Installation Tips

Speakers should be high-quality, high-frequency, and mounted in rigid frames to prevent warping and distortion. The speaker should be able to handle continuous power application with no more than 10 percent distortion. If replacing a speaker, make sure it is the correct impedance for the application.

Speaker wiring should be routed away from alternating current circuits such as certain types of lighting circuits (26 VAC, 400 cycle) and windshield heat inverters. Speakers should not be grounded to the airframe right next to the speaker but should be grounded as close to the audio source as possible to prevent unwanted noise from sneaking in via the airframe. To prevent this possibility, run two wires, one high and one low. Also, make sure that the insulation surrounding the speaker isn't conductive and that the insulation hasn't been packed in too close to the speaker inhibiting cone movement.

Microphone, Headset, ANR Headset, and Intercom Manufacturers

These manufacturers' addresses are listed in the Sources section at the end of this book.

Microphones

Electro-Voice
Flightcom
Telex Communications

Headsets

Acousticom
Aire-Sciences
Audio Communications
Aviation Communications
Aviall
Comtronics
Concept Industries
David Clark
Evolution
Flightcom
McCoy Avionics
MicroCom
Oregon Aero
Peltor
Pilot Avionics
Plantronics
Puritan-Bennett Aero Systems (oxygen mask with mic)
Senheiser Electronics
Sigtronics
Softcom
Sony
Telex Communications

ANR Headsets

- Bose
- David Clark
- Evolution
- Telex
- Intercoms
- Headsets, Inc.
- Audio COMM
- Aviation Communications
- Bendix/King
- David Clark
- Comtronics
- Concept Industries
- Flightcom
- McCoy Avionics
- MicroCommunications
- Pilot Avionics
- PS Engineering
- Radio Systems Technology
- Sigtronics
- Telex Communications

CHAPTER 6

Audio Systems

Audio and Noise

The perception of how well radios work is affected by the amount of unwanted noise and interference you have to listen to while using the radios. This chapter explains some of the sources of unwanted noise and some of the ways good avionics technicians prevent noise and interference using good installation techniques. The pilot plays an integral part in solving these problems by providing detailed documentation about the failure, such as when, where, and how it occurred.

The random noise of Rice Krispies as it meets cold milk is expected by pilots young and old everywhere, but not in their avionics audio. With the advent of modern radios, noise levels actually increased because radios were now more sensitive. With further engineering improvements, avionics radios became quieter, but the problem is still there and much of it will remain until the aircraft's environment changes. The older the plane, the greater the potential for noise haunting the corridors within the wiring and equipment. Old-timers came to expect the inherent noise that was found in early radios, but today, that has changed. Most pilots want the audio in their plane to be as clean as their car stereo.

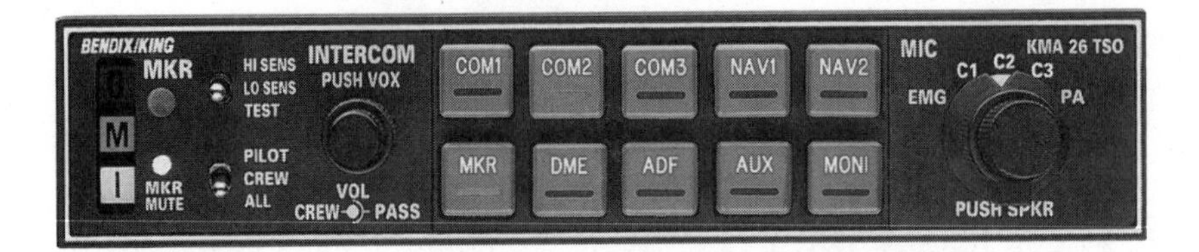

KMA 26 audio panel.

Evolution in nature is something that experts on both sides will argue back and forth, but in avionics there isn't much to argue about. Things change, and that usually isn't painless. However, newer and more sensitive avionics have led to an increase in noise because the radios are able to hone in on the previously undetected noise as well as the weak radio signal. This is amplified and is heard as noise in the headsets and speakers. Sure, there has been some jokes about the squawk being a short between the headsets, but it bears repeating that all squawks should be closely investigated to duplicate the problem. Make sure the pilot has adequately documented what happened prior to the squawk developing. This is accomplished when the technician (or avionics manager) and customer first go over the pilot squawks.

The aircraft is a self-contained, dynamic, plastic and metal vehicle, hurtling across the tarmac and through space, carrying environmentals, instruments, electrical power systems, entertainment equipment, lighting, pneumatics, hydraulics, control surfaces, engines, and avionics. There is something else; something almost alive that we don't like to talk about; something that is so illusive and cunning that it evades its hunters at every turn. You know what it is—noise—you've tracked it, the harbinger of failure.

Knowing about some of the potential noise sources is a giant step in preventing unwanted signals from entering the audio or navigational circuitry. Solutions are usually hindered by a lack of adequate information. The troubleshooter must be fully familiar with the normal operation of the affected system. Communication with the pilot about the failure or noise onset is a good start but must be tempered with insight about the pilot's perception and personal flying habits. What pilots hear may not be what they get.

For example, the pilot may be a weekend warrior, flying for fun and not business. This subtle difference could very well be the distinction in how he/she perceives a given failure. Flying is expensive, and denial is just one way a given noise might be ignored. "Might be just a glitch, will probably go away," exclaims the pilot under his or her breath, as the pilot switches to another radio. On the other hand, the customer could be an aviation professional who is very explicit and concerned about flight problems and will expect all squawks to be equally honored and fixed.

This is the world of audio, the necessary and focused interface between the user and the radio, a very indispensable and complex jungle of dedicated electronic signals that maintain a connection between the pilot and the tower. Audio is the lifeline between the pilot and ground. It is also a distraction from loneliness and boredom. Without audio, there would be total chaos. Planes would collide into each other while confusion on the ground would drastically delay flights.

So what are some of the expected and time-honored problems that creep out of the proverbial woodwork? The onset of more evolved electronics has led to greater sensitivity, and resultantly, noise. Everything can be a culprit. It could be inverters, strobes, alternators, motors, actuators, precipitation static buildup, poor bonding, defective shielding and much more. Finding the source of the noise is the challenge for the hopefully skilled and experienced technician. Specifically, noise is that component of a varying signal with frequencies at levels that are audible, invasive, not direct current, and can usually be traced to such sources as 400 cycles from inverters or 60 cycles from lighting systems. Other potential generators are the spiking signals transmitted over a wide range of frequencies from strobes and bad brushes on generators and alternators.

Noise in its practical form is an annoying squeal, pop, snap, or whine coming from the speakers or headsets. Aside from indicating filter failure or power supply problems, it can affect the pilot's ability to function efficiently after a while. The noise, over a long haul, could easily be enough to impair judgment. The source might be an alternator, inverter, static buildup, defective antennas, poor RF or electrical bonding, defective power supplies, poor shielding, or even equipment failure.

The audio system can be as simple as an internal amplifier in the NAV/COMM unit wired to a common point and switchable from, say, COMM 1 to COMM 2 with mini-switches. Or in the case of a one-radio airplane, the amplifier can be wired directly to the speaker and headphones.

More common, and more complicated, especially in IFR-equipped aircraft, is the panel-mounted amplifier with built-in switching (commonly called an audio panel) or a panel-mounted set of switches connected to a remotely installed amplifier. The job of these amplifiers is to receive low audio (which is basically a weak signal transmitted over many miles), isolate it, send it to the switch-selected output (headphones or speaker), and amplify it so that you can hear what somebody is trying to say to you.

Audio panels tie together multiple inputs to let you use them all efficiently. Instead of having to turn the ADF volume up every time you want to listen to a station, you simply flip the ADF's audio panel switch to either speaker or headphone and listen. This saves wear and tear on volume controls and gives you a central black box from which you can control all your radios' audio. It also can be the creator of the noise and one of the first places to check.

Narco CP136M audio panel.

The easiest unit to install and maintain is the panel-mounted audio panel. Examples of these are King's KMA134 and KMA24, Narco's CP-136, and the former Collins AMR-350. (*Note:* Collins sold its line of Micro Line avionics to S-TEC.)

Audio System Problems

Any kind of switch is always a potential source of trouble, and audio panels have lots of switches. When squawking audio problems, be sure you have the switches set correctly before taking the time and trouble to have someone work on your system. It's all too common for flight instructors to fail to teach students the details of how audio panels are set up. If you rent airplanes, especially those with different avionics setups, take a few minutes to learn the various audio panels so you eliminate lack of knowledge as a source of problems.

In my experience, early audio panels with toggle switches are more reliable than those with push-button switches. Toggle switches are a straightforward design that has proven themselves for many years. Other sources of trouble in audio systems are the switching circuits used to gain the most efficient use of the audio. Without this switching, you'd have to turn down the volume on other pieces of avionics any time you wanted to transmit to prevent the other audio from interfering with your transmission. The other radios are still producing audio, but the audio is prevented from reaching the speaker or headphones by muting circuitry. If the wiring or relays that form the heart of the switching and muting circuits are less than perfect, failures might result, such as audio bleed-through from one system to another.

Following are some common audio problems. Reasons for the failure might involve more than just the audio system, you might see these same problems discussed in other chapters.

1. **Noisy reception.** Check for an antenna problem. Also, it could simply be a weak or noisy received signal.

2. **Distorted or garbled reception.** Check speaker first, then radio, audio amplifier, and associated wiring. If the audio panel uses push-button switches, they could be the problem. Toggle-switch failures are rare. Check for contamination, loose pins, or broken wires on the audio panel mating connector.

3. **Intermittent reception.** This problem can be caused by problems with the radio's power wire, grounds, defective pins on the radio's mating connector, problems with the radio itself, or damaged wiring from recent upholstery work. Check to see if the COMM radio has an intermittent or defective microphone being keyed. This would kill all audio if the microphone key button is depressed or has failed causing a closed microphone key condition. Many technicians test for intermittent audio by turning on the ADF audio and listening to it while proceeding with other tests. If the intermittent audio reception problem occurs, the tech will immediately hear the failure when the ADF audio quits. From there, the tech should be able to isolate the problem further, but at least they will have heard the problem and won't ask you to come back later when it gets worse.

4. **Total audio failure.** Check for the same items as with intermittent reception. But also check to see if you're trying to listen on the speaker with the audio panel set to headphones. Also, make sure the "auto" switch hasn't been moved from where you normally set it. This happens more often than most pilots would like to admit.

5. **Weak reception.** If noise isn't a problem but reception is still weak, check the radio or audio amplifier.

6. **Weak transmission.** This could be from a bad microphone, radio, or antenna.

7. **Modulation, no voice.** Mic jack contacts or the microphone wiring might be defective. Try replacing with a new microphone. Usually, wire breakage occurs directly where it exits the plug or microphone body.

8. **Whine.** A whining sound could be interference from alternator or electric motors (like a motor-driven rotating beacon).
9. **Motorboating.** This is noise in the system that sounds like a motorboat engine. It could be caused by open ground wires, bad grounds, or a miscalibrated radio.
10. **Static.** Static in the audio can result if the shielding between audio wires is touching or from precipitation static on the outer aircraft skin as air, dust, and moisture move rapidly across the fuselage, creating friction buildup. Check that all static wicks are in good shape (they do wear out and even break) and that control surfaces and other hinged surfaces like gear doors are properly bonded to the airframe.

Noise

We would all like our audio systems to be totally noise-free, but we'll always have to put up with a certain amount of interference. Even some of the best-designed and -installed systems will pick up some noise from the atmosphere and from other electrical components in the airplane. You need to know what is normal and what isn't for your particular installation so you don't waste time troubleshooting something that doesn't need fixing.

One simple way of avoiding noise problems is to install avionics of the same brand. Mismatched black boxes can introduce unwanted noise that is difficult to eliminate, although even a full panel of Allied Signal avionics can still cause problems from poor wiring or installation.

Most sources of noise are readily identified but still difficult to track down because of the many routes the noise can take to arrive at the radio and disrupt communications or navigation. Some causes of noise are obvious enough to isolate and reduce to acceptable levels, but others are harder to find and even harder to fix.

How does one deal with these annoying and possibly dangerous phenomena? Even new "glass cockpits" with multiple cathode-ray tube displays in the panel suffer from distortion and interference, plus there are still the irritating hums, whines, and popping noises in the audio.

Knowledge is the answer to improving our ability to search out and destroy radio interference. The aircraft environment is a hotbed of interference from motors, generators, alternators, strobe lights, and poorly installed radios. To make it more complicated, nature has periodically seen fit to blanket the airplane with atmospheric and precipitation static, lightning, and even sunspot activity.

Each of the following potential noise generators can radiate RFI or induce electrical signals directly or indirectly into an avionics radio. The sources of noise are divided into three general groups:

1. Atmospheric static.
2. Precipitation or P-static.
3. Man-made avionics noise (the worst of the bunch).

Don't expect solutions to all of the problem areas mentioned; however, regardless of whether you are a pilot or a technician, you will come away with enough ideas that will allow you to map out an approach to minimize or eliminate unwanted noise signals.

Atmospheric Static

Atmospheric static is caused by RF energy produced by electrical discharges in the atmosphere. Its effects can be effectively controlled through proper use of directional antennas, RF bonding, and internal radio filtering.

Precipitation Static

Precipitation static (also called P-static) results when an aircraft passes through clouds, rain, dust, and other particles suspended in the atmosphere. You might be surprised to know how much dirt is floating around in the air we breathe. Precipitation static is also known as friction charging and occurs when electrons are stripped off the airframe's skin, changing the electrical potential of the airframe. When the static charge discharges suddenly, you can hear precipitation static as a hiss or frying sound in the audio, or see it displayed as nervous navigation needles, or even heard as a blanking of communications.

Proper use of static wicks and good bonding procedures on antennas and airframe grounds can keep precipation static effects within acceptable limits. You can obtain additional protection by ensuring correct routing, shielding, and bonding of wiring harnesses and antennas.

Another important area requiring attention to reduce the effects of precipitation static and lightning strike damage is bonding of control surfaces and other hinged surfaces attached to the airframe. While most control surfaces have a hard metal-to-metal connection to the airframe, static built up on, for example, the right wing might not flow into the right aileron to discharge from the static wicks on the trailing edge of that aileron because of lubricant and dirt buildup on the hinges. You've probably seen the stranded bonding straps that electrically "attach" the aileron to the wing, and you probably were shown to check those during preflight inspections. These bonding

Hinge door bonding.

straps do get frayed, wear out and fall off, so it is important to keep them in good shape.

It's important to note there should be two or more hinge bonding straps; if one has broken, don't ignore it. The one that is broken might just be the one that keeps the wing from having a hole blown in it, or prevents the strike from traveling into the plane and burning a hole in your leg or arm. Problems with bonding straps can crop up after paint jobs. If the control surfaces were removed for painting and rebalancing (which conscientious paint shops will do as a normal part of the job), someone might either leave the bonding strap off completely or reattach it on top of the fresh paint, instead of removing the paint where the strap attaches to ensure a good metal-to-metal bond. There should never be less than two electrical bonding straps, especially on control surfaces.

For long-term protection, do not ignore the condition of your airplane's static wicks. These devices are designed to discharge the electrical potential built up by flying through precipitation static-generating atmosphere, and their effectiveness diminishes over time. Next time, before you go flying, look at the static wicks. The discharging occurs at the tail end of the wick, and if the tail is missing a lot of its brushy ends or is broken off completely, it can't do its job. Another thing, just because they look okay doesn't mean they are. If you look on the underbelly of some static wicks you will find a small resistor. This could be open and therefore totally ineffective. Many are installed over painted surfaces and might as well be left off for all the good they do. Unfortunately, many mechanics ignore static wicks during routine inspections and never call for their replacement. Keeping static wicks in good shape is especially important on some of the fast, modern, single-engine aircraft like Piper's Malibu and the turbine-powered TBM 700.

A good example of a bonding problem occurred to one of my customers in southern California. Many pilots and technicians aren't aware that a major cause of precipitation static is dust, not just moisture like snow and rain. Range on this customer's NAV receivers was a mere 30 to 40 miles, well below what it should be. Rather than try one thing and test fly it and then try another repair and another test flight, we decided to shotgun the airframe's bonding by rebonding all antennas, control surfaces, and hinged surfaces like gear doors. This decision was made based on the fact that it was obvious that the antennas had not been touched for some time. The results were outstanding: NAV range more than doubled to over 100 miles. At the time, there was a severe drought in California, which resulted in a lot of dust suspended in the atmosphere, even at higher altitudes. (We didn't jump at the bonding problem initially; the first step in a NAV problem is always to recalibrate the units as close as possible to factory specifications before trying another more expensive fix.)

A good avionics person should be able to take one look at the antennas and determine if they should be removed and rebonded. All the obvious signs are there. Warning flags such as the missing sealer around the base of the antenna, to the gaps under the antenna base and even corrosion, are just a few factors that should be given serious attention.

Thunderstorms can create conditions that generate emissions from the extremities of an aircraft, commonly referred to as "Saint Elmo's fire." These emissions are displayed as corona and streamer discharges and usually last for only a few minutes. Streamers can grow from 20 to 30 feet long but usually don't contribute to significant navigational difficulties unless antennas aren't well bonded to the airframe. Saint Elmo's fire was named for Saint Peter Gonzalez (don't ask me how you get Elmo from Peter Gonzalez) who lived in the thirteenth century and was named the patron saint of Mediterranean sailors. The discharge is said to be a sign of Saint Elmo's protection, so I suppose pilots can take it as such, too, being sailors of the atmosphere.

Another factor that can contribute to airframe charging is exhaust from the engines. Positive or negative ions (atoms having a positive or negative charge) are selectively drawn to the exterior cowling by the exhaust, creating either a negative or positive charge that migrates to the airframe. The amount of this charging is directly related to throttle settings, and the diameter and length of the exhaust stack.

Once part of the airframe develops an electrical potential, either from friction, thunderstorms, or engine charging, interference or noise can be generated three ways:

1. Corona
2. Streamer discharge
3. Sparkover from surfaces that aren't RF or electrically bonded to each other

Corona discharges cover the RF spectrum from very-low frequency (VLF) to high frequency (HF) and can significantly affect very-high frequency (VHF) and ultra-high frequency (UHF) communications used by avionics. Aircraft antennas are an ideal example of extremities, protuberance, or sharp points that have the potential to discharge the charge collected on the airframe surface.

Here you have an entire avionics system, complete from antenna to radio, subjected to considerably more current than it was ever designed to handle. Large portions of the charge are discharged harmlessly into the atmosphere, but a small, tenacious, and sometimes lethal dosage can reach the radio. Even if not damaged, the radio most likely will shut down because of the RF and IF stages being driven into oscillation and saturation. This suspicion can sometimes be confirmed by inspecting for small, needle-like black dots on the surface of the antenna. These penetration points are a telltale sign that possible damage has occurred. Note that this damage can also be caused by small tendrils of charges from lightning strikes. Discharges from control surfaces, nose cones, wing tips, and other unbonded surfaces can corrupt radio reception or transmission on aircraft smaller than light jets because the antennas are in close proximity to the unbonded surfaces. If you suspect that lightning has struck your plane, have someone qualified in lightning strike damage inspect for damage. The radios may have sustained a tendril bleed-off from the main strike, which degraded the radios to the point of reducing their potential life span. When you make your claim, request that all the radios be bench tested for degradation. Additionally, aside from what prompted the investigation, other parts of the wiring and airframe should be inspected for any obvious signs of pinholes or burnt spots.

Plastic windshields (acrylic) wing tips, vertical fin caps, and other plastic fairings are potential sources of streamer discharge. During flight, the windshield, for example, builds a charge caused by frictional interface with dust, moisture, or precipitation, and the charge continues to build until it jumps across to the metal airframe. It is very important to make sure the windshield is electrically bonded to the aircraft structure, not suspended in sealer, otherwise there could be catastrophic failure of the plastic as it is overheated. Once the streamer begins, it will continue as long as the charge is maintained. Few people are aware that some waxes are not for use on aircraft and are prone to streamer production. I would recommend you check this with the wax manufacturer or stop using a particular wax if you suspect that it's causing a streamer problem.

TIPS ON PREVENTING PRECIPITATION STATIC

Follow these guidelines to prevent precipitation static buildup.

1. Use proper antenna spacing as recommended by the manufacturer.

2. Ensure that the antenna base is properly bonded to the airframe skin. That means metal-to-metal contact, with no paint or other insulating substances between the base of the antenna and the skin of the aircraft. The product called Ultra Seal Dayton Granger is highly recommended for bonding antennas to the aircraft skin. Be sure to follow the instructions provided by the manufacturer. There are other alternative sources, but you will have to contact the writer for these.

3. Antennas usually shouldn't be installed within three feet of the windshield, although it is frequently easier for installers to do so. It is more difficult to install an antenna a proper distance from the windshield because access to the airframe skin is harder, plus antenna cables have a longer run to the radios, but the extra effort will be worthwhile. The expense of relocating a

badly placed antenna will exceed the initial cost to do it right the first time.

4. As a general rule, you can minimize precipitation static noise problems by making sure there is metal-to-metal contact between all access doors, antennas, and onboard equipment, whether avionics or electrical. If not installed, bonding straps should be added to control surfaces. Although this may be very difficult, the chances of noise problems occurring will be much lower if there is less than .25 ohm resistance between the control surface or gear door and the airframe.
5. Don't ignore inspection panels as a potential source of noise, too. Panels need to be bonded to the airframe as well. Usually, the panels are metal and are bonded through normal installation. Metal screws attach the panel to the metal airframe. If there isn't bare metal-to-metal contact, however, the panel isn't bonded and could be a source of noise.
6. Radio racks should be bonded with at least two, 1-inch-wide by 5-inch-long, cadmium-plated brass straps connected between the radio rack and airframe ground. If possible, also have the mounting base of the rack RF bonded; this will doubly improve the bonding and ensure the installation is free from the troubling effects of extraneous RF interference.

This doesn't seem to be complicated, does it? Just throw on a few bonding straps, check this and that, and that solves your problems. The truth is, these considerations are frequently ignored in the field. Most of the aircraft I've evaluated over the last two years were in dire need of the techniques explained above. Avionics and maintenance shop managers need to advise their customers about the necessity for good RF bonding. It is something that should be done during a normal installation, checked as part of a routine inspection, and should not be ignored.

Man-made Noise

Being familiar with noise sources is a giant step toward preventing unwanted signals from entering the audio or causing navigation signal problems. The problems that arise are frequently the result of lack of adequate information. Attempting to troubleshoot a problem without being prepared might result in a technician setting off in the wrong direction and wasting time and money and not fixing the original problem.

There are many sources of man-made noise. The following explains many of these sources and what can be done to minimize or eliminate them.

Wiring is one of the weak components of avionics systems and can be divided into four categories:

1. AC (alternating current) power
2. DC (direct current) power
3. Pulse circuits (transponder and DME)
4. Sensitive circuits (autopilot feedback circuits)

Interference or noise signals can be divided into three categories:

1. Electrical fields
2. Magnetic fields
3. RF (radio frequency) fields

These signals, in turn, can be injected into potentially sensitive areas of an avionics system via antennas, power lines (wires), or signal lines.

There are four coupling modes you should know about:

1. Conductive
2. Capacitive
3. Inductive
4. Electromagnetic (RF)

Coupling, the introduction of unwanted signals or noise into electric and electronic circuits, occurs in about 88 percent of all aircraft wiring. For the most part, coupling effects are neutralized by circuit design, wire separation, and other rigorous installation practices. Power lines, AC/DC relays, and solenoid lines are key concerns for coupling interference into analog, digital, and audio signal lines. Conductive, inductive, and capacitive coupling take place via magnetic fields cutting across equipment and wiring. Electromagnetic coupling is RF entering through antennas.

A good example of a simple noise problem is the complaint of noise in the speaker from one of the radios. This could be caused by low voltage. Most radios are designed to operate within specified input voltages. A 12-volt aircraft, for example, might have radios that specify input voltage from 11 to 14 volts. Low voltage could prevent the radio's circuitry from operating efficiently and could allow the generation of noise internally or into the audio system.

To check for this, you can simulate a low-voltage condition by testing your radio's reception on the ground with the avionics on but without the engine running. As you turn on more and more electric items, the bus voltage will drop below the battery's normal 12- or 28-volt system. As the voltage drops, you may hear noise in the speaker. You can measure the voltage drop by simply checking the bus voltage with a voltmeter attached to the bus bar under the instrument panel, where all the circuit breakers are attached.

If you are getting noise from your speaker, then make sure the alternator or generator is properly charging the electrical system. If the system is charging properly (usually 2 volts above system voltage for a 12-volt system and 4 volts above for a 24-volt system), then you won't waste time chasing a noise problem that is caused by low voltage. Keep in mind that if troubleshooting a noise problem using battery power alone, you can easily drop the battery below its nominal voltage during testing. Always use a strong, fully charged battery for testing, or better yet, use an external power source such as an APU or the new Start Pacs from Hawkins Associates

Company, Inc. to eliminate low voltage as a cause of a radio problem.

There are times when the radio's power supply will suffer a component failure and work fine with the normal operating voltage applied but act up when the alternator is off-line. It might be embarrassing to find that you wasted several hours troubleshooting a problem that was simply a result of a defective power supply or low voltage. Power converters that convert 12 volts to 24 volts have been known to cause this type of problem.

Relays and solenoids have been sources of radio noise in the past, but this hasn't been a problem with modern aircraft. The problem lies in the older planes that are still in operation. If an undesirable noise occurs only when you flick a switch or activate a certain circuit, then you can trace the noise to a worn relay or switch with burnt contacts. The spark created by the opening and closing of the switch contacts is most likely entering the audio through the wiring attached to the switch, like an antenna.

Inductors, such as solenoids and relay coils, tend to create high voltage transients when their magnetic fields collapse as the circuit feeding the coil is opened. The magnetic lines of flux in the collapsing field will induce a high voltage back into the DC relay coil, which then races back into the system wiring, possibly causing damage to solid-state equipment or noise in the audio. To prevent this, a diode is installed across the coil of the relay. Polarity is very important; only a skilled technician should install the diode. This prevents the voltage produced by the collapsing field to radiate to surrounding wiring, and tends to increase the life of the coil.

Strobe lights are frequent sources of noise complaints, causing loud popping sounds in the marker beacon, navigation, and ADF audio. Because strobes are RF generators by design, it's no wonder thousands of aircraft are plagued by strobe noise squawks. The noise can be transmitted directly to nearby antennas or induced into electrical or radio cables routed too closely to strobe power supplies or strobe power wires. Shielding cables and wires or filtering strobe power supplies can eliminate most of these problems, and so can

Wing tip strobe. These usually are not as much a problem to the avionics as the tail strobe light or the power supply.

simply relocating the antenna, the strobe power supply, or in some cases replacing the strobe light with an attenuated version to reduce the radiated interference that emanates from the light.

If the power supply is causing the noise problem, replacing the power supply or installing additional filters might be the only solution. Most strobe manufacturers will overhaul strobe power supplies for a reasonable price, so the simplest solution might be to box up the power supply and send it to the manufacturer. Ask the company to let you know if the condition of the power supply could have caused your noise problem. Of course, the answer may be helpful if the problem is still present after reinstallation.

Audio Noise Filters

There are four kinds of filters:

1. Low-pass
2. High-pass

3. Bandpass
4. Band elimination

Most filters used in noise suppression are the low-pass variety. Filters suppress noise by shunting the undesired energy or noise through a capacitor to ground. To protect the capacitor's dielectric material, the capacitor must have a voltage of at least twice that of the circuit voltage circuit to which it is connected. If electrolytic-type capacitors are used, use a voltage rating of four times the circuit voltage because of potential dielectric failure.

Don't go overboard with filter capacitance, though. Capacitance is measured in microfarads (mf), and although a filter with 1 mf capacitance is ten times as effective as a 0.1 mf capacitor in practice, the small capacitor is more effective at higher frequencies. So it's important to match the filter to the frequency of the unwanted signal you're attempting to eliminate.

The most frequently used filters are low-pass filters with cutoff frequencies in the region of 1 to 10 kilohertz. They have a very small insertion loss for DC circuits and ordinary power-frequency currents, and they can aggressively attenuate all radio interference currents.

One relatively simple way to locate the course of the noise and determine the purity of the DC feed and return lines is with an oscilloscope. Every noise has its own unique signature. Keep in mind that the technician might not want to use a filter for the exact value that is a direct computation of the noise trace as indicated on the oscilloscope. The technician needs to attenuate the offending frequency, and that could lie at a point slightly above or below the center of the noise's frequency range observed on the scope. Many factors affect the frequency transmitted by the source of the noise such as length of wires, capacitive effect caused by wire routing, and engine RPM if the noise is related to the generator or alternator. Experimentation, such as trial and error may be needed to find a filter to eliminate a specific noise.

Selecting a filter can be a difficult choice, but not if there are only a couple on the market. Lonestar in Texas has a 10 amp, inline filter specifically for avionics. It is PMA'd and is quite popular and the only one I would recommend at this time.

Finding and Eliminating Noise

Few avionics shops have elaborate signal analyzers, so process of elimination is the only inexpensive and practical way to locate positively the source of an offending noise. Begin by positioning the airplane well away from fluorescent lights, power transformers, operating cars, and other aircraft.

Start the engines, turn on the avionics and other electrical items, and listen to the headset or speaker for the offending noise. If you can't hear it, operate the controls in the aircraft as if you were flying to see if control surfaces or cables are causing the noise. Turn on the autopilot and run it through its ground tests to see if it is making the noise. Continue by running the engines up to maximum RPM, then vary the RPM up and down, slowly, while listening for the elusive suspect. Operate the cowl flaps, fuel pumps, and other intermittently used equipment. If you need to test fly the airplane to find the noise, try various combinations of equipment used while flying to bring out the conditions that lead to the noise generation.

Some of the most often ignored sources of noise generation are improper bonding or grounding of equipment chassis or racks, shielding terminations, common ground tie points, and resistance/inductive effects.

Here are some tips to prevent or eliminate noise problems.

1. Reduce the possibility of interference coupling by making sure the rack holding the radio and the radio itself are at the same or close to the same potential. Grounds from avionics equipment should have the same ground potential as other radios in the aircraft, whenever possible.
2. Shielded wiring between the radios should have the shielding terminated no further than 1.5 inches from the connector, and only at one end of the wire. The attaching shielding pigtail must be less than 6 inches long. In order to work effectively,

the shielded wires must be bonded only at one end, and all the strands must be included. This should be done at the source of the signal with as short as possible. The purpose is to avoid capacitive coupled-ground feedback loops.

3. Make sure all wiring is capable of carrying the current rating of the equipment to which it is attached.
4. Transient or spike frequencies are usually found in the 10-Hz frequency range, while continuous noise interference is usually greater than the 10-Hz range. Select filters based on this data.
5. Digital circuits or lines, which are found more frequently now in advanced avionics equipment, are very sensitive to interference and should be made of twisted-pair or twisted-shielded-pair wiring. Signal wiring should be used with twisted pairs or with shielded balanced coaxial lines. Any induced voltage (noise) in one wire tends to cancel out any oppositely induced voltage in the parallel twisted wire. The only deviation from this approach is when the wires are high energy and run over long lengths. Grounding both ends may be the only way to prevent interference levels by lowering the impedance of the shield to ground. Where to place these grounds must be determined by trial and error.
6. When power-line impedances are in the 10-ohm range and the digital and audio impedances are in the 1000-ohm range, the power circuit conductors can be unshielded and audio line conductors shielded.
7. Ask the technician to make sure low-voltage lines are not running close to high-voltage lines and communication lines are not too close to DC or AC power lines.
8. Equipment chassis and shields should never be used as signal return path.
9. Signal wires must not be routed or bundles with hard current cables such as motors, actuators, servos, or other power lines.

In most cases, a one-to-five ratio should be used on all bonding straps for RF cancellation. These should be made of flat brass or beryllium/brass stock. The combination of certain lengths of non-RF-type bonding straps such as braided cables, or simply wires and the capacitance caused by other metal sections of the installation, can create circumstances that will produce resonant RF radiation in the 50 to 500 MHz range. This range covers the frequencies used for communication and navigation and thus could cause just the type of interference you don't need.

If you have consistent static in the speaker, it could be caused by a failure of the squelch circuit, or the squelch might not have been set to a muting position. You'll always hear noise in a radio from all kinds of man-made and natural sources. The squelch circuit is designed to mute out the static until a strong clear signal—the signal that you want to hear—is received.

Signals such as those between the autopilot computer and servos and flight director should not be wired to travel through the shielding of the interfacing wires. Never allow equipment racks or their chassis and the attaching harness shields to be used for signal returns. For proper delivery of signals between equipment, use shielded, twisted-pair wires with no less than six twists per inch, or use balanced coaxial lines. This will ensure than any unwanted voltages generated (noise) will cancel each other out. The induced voltage from one wire into another will be 180 degrees out of phase and is therefore theoretically neutralized.

When using high- and low-signal wires in an avionics system, treat them like oil and water: They don't mix. Make sure they are routed and tied out separately. To reiterate: Each signal line should have its own return line, independent of any other, running as close as possible to the signal line, preferably as a shielded twisted pair of wires. The shielded twisted pair fits the bill and is the best choice.

Shielding is effective only against capacitive-coupled noise sources. These types of noise are produced by power sources like alternators or even inverters (which change voltage from DC to AC, like 12 volts DC to 110 volts AC. Before searching for shielding problems, take a close look for poor grounding of the alternator or a bad alternator filter (the capacitor attached to the alternator). The ADF can be shut down completely by a defective alternator diode; this can generate a frequency close to one of the local radio stations. Crossing power wires at 90 degrees will reduce the effects of capacitive coupling. Also, distance between cabling will dramatically reduce coupling, both capacitive and magnetic. The distance doesn't have to be a great amount, even as much as half an inch can virtually eliminate most of the coupling effects.

You can separate power cables yourself easily using high-quality nylon ties obtainable from any electronics store, or even Home Depot. One method is to make a figure-eight around the two cables, then secure the intersection of the eight with a second tie. A better method is to cut a small half-inch piece of rubber tubing and place it between the two cables. With a nylon tie, form a single loop around the first cable, then bring the end of the tie through the tube, around the second cable, and back through the tube. Now insert the end of the tie into the locking part of the tie and pull snug. You now have a half-inch standoff securely holding the two cables apart.

Differences in potential between components or between the instrument panel and instruments are significant targets for noise or adverse signal generation. For example, if an electric gyro should start up and there is a voltage potential between the panel and the gyro, a momentary or even continuous voltage will exist. To check this, hook up your voltmeter to read system voltage on the electrical bus. With the engine running and all electrical equipment and radios on, move the selector on your voltmeter to the lowest voltage range until the lowest reading is showing on the meter. This voltage is your interference. Now you can turn off each piece of electrical or avionics equipment and see which one makes the low voltage go away. That equipment is your culprit. If you don't see any voltage on the voltmeter except for system voltage, then leave the setting on system voltage and monitor the needle carefully (or number, if it's a digital readout) for sudden surges as you turn off and on each piece of equipment. A surge will indicate a potential noise-generating piece of equipment. If the panel shares its common ground with another audio component or the microphone circuits, a direct coupling could take place, and it will have to be tracked down if you don't want to have to put up with the interference.

If you suspect inadequate bonding at a given radio, check with a VOM designed to measure below 1 ohm. You'll most likely find primer or corrosion between the radio rack or tray and where they attach to the panel's structure. Just cleaning the mating surfaces isn't enough—you must corrosion-proof the surface as well. Use Alodine 1200 solution for aluminum, which is the material you're most likely to find in aircraft avionics installations.

Many technicians encounter autopilot interference problems because of poorly bonded equipment racks to airframe grounds. Routing heavy current-carrying cables next to signal wires can induce random spikes that can make for a frenzied situation for the pilot, especially if it happens on a climb using the autopilot. The airplane can lurch either up or down, right or left, suddenly and momentarily, and sometimes for longer durations, possibly causing loss of control. Finding this problem can be easy or quite difficult, depending on the autopilot or the airplane. Isolation of the problem is easier if the source of the problem is caused by, say, keying of the mic or operation of some other piece of avionics. Here are some ideas that the technician should consider when troubleshooting autopilot interference problems:

1. Place a filter on the generating source of interference.
2. Reroute heavy-current leads.
3. Bond autopilot and other avionics computer racks properly.
4. Bond servo properly.
5. Shorten ground leads.
6. Reroute communications coax cables away from autopilot signal wiring.
7. Relocate any offending antennas.

8. Bond antennas to airframe properly.
9. Relocate autopilot computer from source of interference.
10. Bond any surrounding access covers (near autopilot components) that might be building up static charges.

SIX TROUBLESHOOTING STEPS

1. **Recognize symptoms.** Visually, olfactorily, or audibly inspect; look for any deviations from normal operation.
2. **Gather symptom information.** Find out as much about what is occurring as possible. Check for the "when, how, where, how high/low, etc."
3. **Make a list of potential component faults.** Create an operation logic flowchart and applicable error locations.
4. **Narrow the source** of the problem down to a system.
5. **Localize the failure** to circuit within the system.
6. **Analyze the circuit.** Note the cause of the failure, what requires replacement, and what should be replaced as preventive maintenance.

Exposing contributors to the noise dilemma is half the battle of calming the torrent of noise that floods modern avionics equipment Poor troubleshooting without good solid problem-solving tactics can be expensive. Strapping on a fully loaded arsenal of knowledge before simply replacing components will enhance the odds of success.

Audio Panel Manufacturers

Some audio panel manufacturers are listed below. See Sources for addresses.

Bendix/King

Narco

Radio Systems Technology

S-TEC

Sigma-Tek

CHAPTER 7

Antennas

This chapter covers antenna polarization, how to choose an antenna, plus troubleshooting and rebonding of antennas.

Antenna Polarization

Ever wonder why NAV antennas are horizontally aligned with the ground, and COMM antennas point vertically? It's because the radio signals are polarized; an analogy would like polarized sunglasses where only certain types of light are let through the lens. NAV signals are polarized so that the signal that reaches the receiving NAV antenna is horizontally polarized, or parallel to the earth's surface. The antenna's horizontal mounting makes it able to better pick up the horizontally polarized signals.

The vertically polarized signals that travel to the communications antenna are "ignored" by the horizontally positioned NAV antenna but picked up by the vertically mounted COMM antenna.

Choosing an Antenna

The best antenna for each piece of avionics on the aircraft is usually the antenna recommended by the avionics manufacturer. Don't shortchange on antennas when having new radios installed, because without a good antenna, expensive radios will be useless. Some aircraft owners have actually tried to use their second ADF antenna for a Loran; don't do it, even though theoretically it should work. More details on antenna are given in the chapters on each type of equipment.

Antenna Problems

Checking the aircraft's antenna system is fairly easy and should be done on a regular basis to prevent unscheduled failures. Have a shop place a RF (radio frequency) wattmeter in-line of the coaxial cable going to the antenna, and rotate the slug to indicate flow toward the antenna. Select a test frequency. Key the microphone and begin to speak. The wattmeter should indicate power out (to the antenna) and modulation.

If power out is displayed on the meter, the next step is determining if the carrier wave can be modulated when you talk. Blow or whistle into the microphone and observe the power out reading for rapid variations. This will give you a good idea that the transmitter should be transmitting the carrier wave and voice modulation. If there is no sign of power out and the radio was working before testing, a connection error has most likely occurred or a connector has failed.

Radio transmissions are electromagnetic fields, and they have two field components that are important to antennas. One is the H-field or magnetic component. The other is the E-field or electrostatic/capacitive element. H-field antennas are designed to receive magnetic radio signals that are less susceptible to precipitation static interference, however by their very design, they are prone to noise effects from 400-cycle AC. H-field antennas, designed with a coil wound on a ferrite core, are very efficient for receiving radio signals, but also induced alternating current. Horizontally polarized antennas such as the navigation balanced loop antenna and Omega system antennas are H-field types.

A combined loop/sense antenna has an H-field component, which is the loop portion, and an E-field component, which is the sense antenna. Vertically polarized antennas such as communications, loran, DME, and transponder antennas are E-field types and are designed to receive electrostatic/capacitive radio waves. These antennas are usually configured to form either a plate or a rod shape. Look at any communications antenna on a Beech, Cessna, or Piper single and you'll see either a blade or a rod antenna mounted on the top or bottom of the fuselage. These are E-field antennas. Most of the towel-bar navigation antennas are balanced loop H-field antennas; however, the V-shaped rod or gullwing style NAV antennas found on vertical stabilizers are E-field antennas and should be replaced whenever possible.

Antenna systems are often consistent, but unnecessary, causes of avionics failures. Problems caused by antennas can include intermittent transmission and reception, hum, distortion, or breakup. As with most avionics troubleshooting, start the process with simple items like the microphone.

Gullwing NAV/COMM antenna.

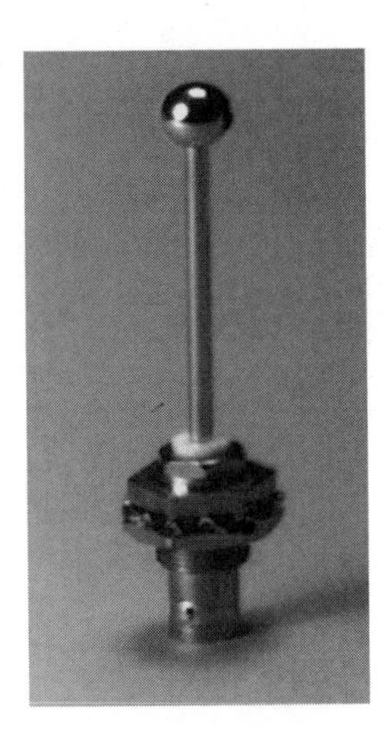

DME/Transponder rod antenna.

RF coupling can be caused by a defective antenna, coaxial cable, or poor bonding of the antenna. Each possible source of trouble must be eliminated until the problem is solved. After bonding the antenna, did the transmission or reception improve?

Inspect the antenna. Look for small but obvious cracks on the surface of the antenna—there could be hundreds. These cracks admit moisture, destroying the ability of the antenna to function within its design parameters. Replacement is the only option as field repairs are neither practical nor recommended. The damage is already done, with moisture having found a home in some secure and dark cavity. Some shops sand the antenna, fill the cracks, and repaint. This isn't a good idea, because paint will most likely reduce the effectiveness even more, and in any case, the internal damage has already occurred, so you are simply covering it up and sealing it in. Besides, if moisture has entered through the cracks and then they are filled, the moisture is now trapped. Freezing will expand the moisture and cause further damage.

Rebonding Antennas

The first step to RF rebonding is to remove the antenna. You might need someone else inside the aircraft to hold the nuts for you, unless nutplates are installed. If so, cautiously back the screws out; it is possible to twist the nutplates off. Be very careful when disconnecting the coaxial cable from the antenna; coax connectors can be very sensitive to mishandling. First use your fingers; if too tight, then apply gentle counterclockwise torque using a wrench or small pliers.

Blade antennas may have a cork or rubber style gasket between the base of the antenna and the fuselage. Discard the gasket and use the procedure described below. Unless the gasket is made from a metallic embedded material, it serves no purpose. Besides, even the metallic gaskets must be replaced with a new one. The antenna *must* be metal-to-metal fastened to the aircraft skin, unless part of a storm detector system like the Stormscope® or Strike Finder®. With these antennas, there is no metal-to-metal contact. Static dissipation is accomplished with a special coating on the antenna that discharges the static off to the aircraft skin.

Clean the metal base of the antenna with acetone to remove adhesive residue from the old gasket. Using a Scotchbrite™ pad, scuff the metal base until it's shiny and clean, but no more. Clean the mating footprint until it is also shiny, but do not remove metal.

Apply Alodine 1200™ solution using Scotchbrite™ (wear rubber gloves while doing this), wait two to three minutes, and remove the excess by dabbing gently with a sponge or soft cloth, but don't rub it off. The Alodine 1200™ is soft and can be easily scrubbed off. Again, make sure the contact area on the airplane's skin is cleaned and prepped using the same technique and the Alodine 1200™ as you did with the antenna base. Allow to dry for no longer than 20 to 30 minutes. This is important; Alodine 1200™ becomes nonconductive after a period of time; from experience, 20 minutes is a safe time frame.

(*Note:* Alodining is the chemical application of a protective chromate conversion coating on aluminum with low copper alloy content to prevent corrosion. Be sure to check with the local sewage treatment facility about the disposal of used Alodine. It doesn't take much of any chemical, including the acetone used to clean the affected area, to violate of EPA regulations.)

Reinstall the antenna within 20 minutes. If the base of the antenna is provided with raised mating surfaces, the only areas that need bonding are these points, which should solidly make contact with the airplane skin. Sometimes you have to use spacers (preferably a tapered or contoured shim) to bond the antenna to a curved surface. If so, make sure the spacers are conductive and electrolytically compatible with both antenna and aircraft skin surfaces.

At this point, it should be mentioned that there is a much better method of bonding antennas. A well known antenna manufacturing

Alodine 1200™ recently mixed.

company in Florida, Dayton Granger, has developed a product called Ultra Seal. This product solves several antenna-related problems: There is no need to apply Alodine 1200™, plus it provides a much improved pressure seal, the sealer does not blow off during flight and, finally, the RF bond is far superior with radiation patterns being relatively equal in all directions. One more plus, the sealer is easy to remove and the need to R&R (remove and replace) the antennas is eliminated. Recent tests show that antenna installation is almost a lifetime installation. Lightning strike tests by Dayton Granger were completely successful in demonstrating that this system is highly effective in processing high current discharges with little or no damage.

Composite airplanes and their control surfaces, gear doors, etc. are bonded essentially the same as metal airplanes. The essential difference is the conductivity of the composite material. Graphite is somewhat conductive, whereas Fiberglas is not conductive at all. In cases where the airframe is totally nonconductive, conductive paint is available and should be applied. The manufacturer may have embedded a wire mesh or screen-like product throughout the skin to protect against lightning strike. (This screen will help in providing a ground plane if the antenna can be mated directly to the mesh.) In addition, a metal ground plane (aluminum plate) for the antenna might have to be installed, bolted to the mesh so future removals won't require repeated cleaning of the mesh. The antenna would then be mounted to the metal plate.

The process of installing or replacing an antenna demands that the mechanic, installer, or technician use the same time proven application processes to clean, corrosion proof and reinstall the antenna, each and every time. There can be no shortcuts, except for the Dayton Granger Ultra Seal system. Remember that the antenna system is the pilot's life line to the outside world and short cuts can be disastrous.

CHAPTER 8

Communication Receivers/Transmitters

Most pilots probably think COMM radio transmitters and receivers are the most important avionics equipment in the aircraft. After all, when becoming a pilot, the very first radio you used was most likely the communications radio. If you're like most of us, you were probably nervous, worried that you weren't going to do it correctly and that if you didn't, your voice wouldn't be converted into electrical signals by the microphone, amplified in the communications radio, and transmitted via the antenna to anyone on the same frequency. After that, it is too late to take back what was said and now you are doubly tense.

Communication radios will always be important, even with the advent of up/down link of data to onboard display screens. They are frequently your only contact with the outside world, especially when flying inside dank, gloomy clouds. Transmitters also use a lot more power than most avionics, and they are the first to go when you lose your alternator and end up draining the battery. It never ceases to amaze me how many "I learned about flying" stories you read where pilots don't notice the alternator is dead until they start to wonder why nobody seems to hear their transmissions anymore. Having a reliable communication system in your airplane makes your airplane

Communication radio KY-196B.

more useful and provides a greater degree of mental and physical security.

The communications transmitter amplifies the electrical signals from the microphone and delivers the amplified signals to the communications antenna, where they are transmitted as radio waves. The reverse occurs when the receiver amplifies the weak signals received at the antenna to the voice we hear on the headphones or speaker.

Aircraft communications radios transmit amplitude modulated (AM) signals in the VHF range of the radio spectrum. Many people assume that because COMM radios transmit on frequencies from 118 to 136 MHz that the signals are FM, or frequency modulated, but that isn't the case. AM allows more frequencies to be used in a particular range of the spectrum, and if FM were used for communications radios, we wouldn't be able to fit as many frequencies between 118 and 136 MHz. Also, although it provides better sound quality, FM is more complex and thus more expensive than AM. For ordinary voice transmission, we simply don't need the quality FM offers; we'd rather have the greater number of available frequencies and lower-quality audio. I'm sure satellite communications will eventually replace the AM, someday, but that day is not yet now.

FCC Frequency Change

The use of older VHF aircraft radios not meeting a frequency tolerance of 0.003 percent was established to be illegal to transmit by the Federal Communication Commission on January 1, 1997. Transmitting on these radios is illegal; however, receiving radio frequencies is obviously okay. Frequency tolerances for the respective radios can be found in the user's manual or maintenance manual. Some models can be upgraded, modified or replaced to meet the new FCC standard. Two or three companies have filled the vacancy, such as Narco, McCoy, and TKM. In one form or another, they can replace the older radios with modern, digital units that provide reliable operation, if installed properly.

Previously, the standard was 0.005 percent. Basically what the tolerance means is that a transmitter must transmit on the frequency selected within the tolerance standard. If the transmitter frequency drifts off beyond the tolerance limits, it could overlap and transmit on other nearby frequencies, causing no end of confusion in the aviation spectrum. With most radios using the full 720 channels allowed by 25 kHz frequency spacing, keeping transmitters within the allowable tolerance is even more important.

Installation Tips

Modern, panel-mounted radios are highly sensitive and efficient. Mixing them, however, in a hybrid fashion, might not only cause cosmetic problems, but it could also create physical and electronic

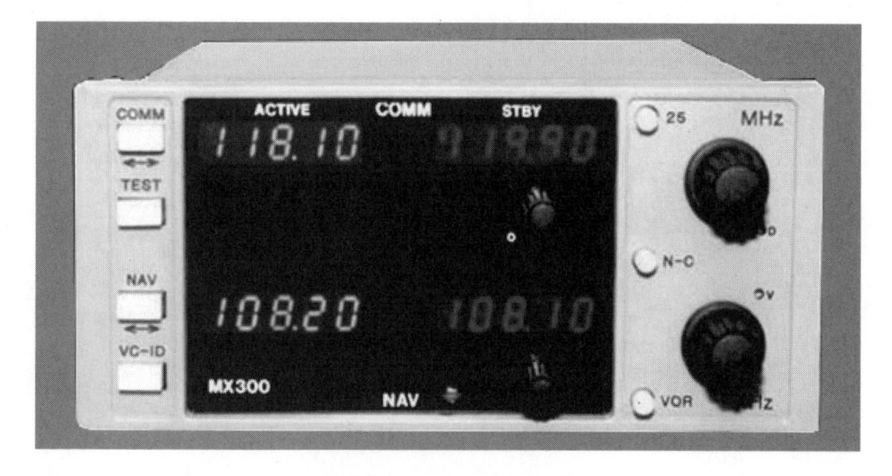

MX 300 NAV/COMM replaces obsolete radios affected by new FCC standard.

incompatibility problems between radios, increasing installation and final checkout costs. I recommend settling on one brand for the whole avionics suite. If the customer can't afford the entire suite at once, assemble the harness complete, but have the radios installed piecemeal, with an eye towards ending up with a complete installation. If you are the pilot or aircraft owner, make sure you get a written assurance that the future radios will fit and not require extensive rework. The installer should be able to easily predict how well the radios will fit without having all of them available. If not, he or she shouldn't be installing avionics.

Selecting wire for the harness and coaxial cable for the antenna system is the responsibility of the shop doing the installation, so make sure the installers are using the highest-quality material. Careful attention to detail in the attachment of brackets, angles, fasteners, wiring harness assembly, RF and electrical bonding, and structural integrity will help you end up with a reliable avionics installation that will withstand rigorous use in the aeronautical environment.

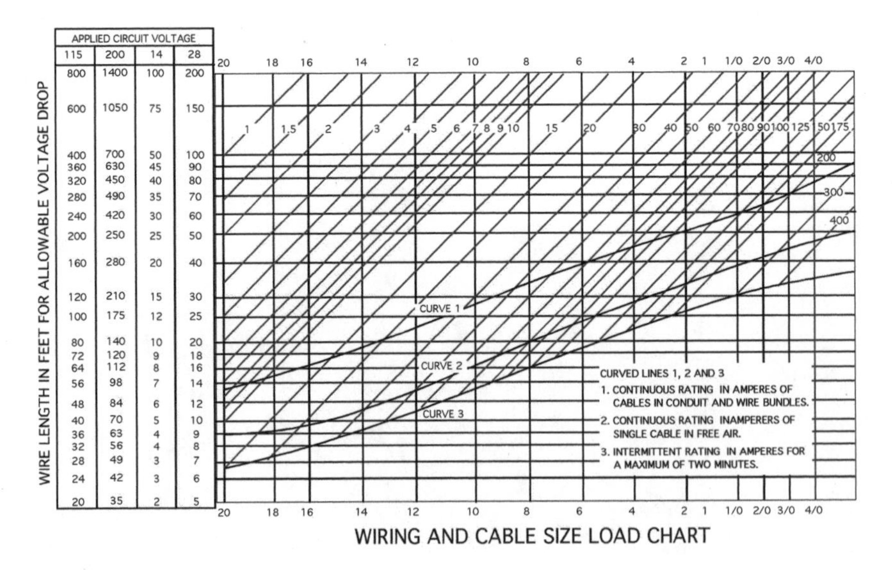

Wire/cable size chart.

When installing a new radio, use the antenna recommended by the radio manufacturer. A mismatched antenna and radio may result in reduced transmission and reception range.

If this is a retrofit radio, the older the wiring, the higher the possibility of damage from nicks, chafing, and cracking of insulation. Consider having new wiring installed, and new circuit breakers, too, as they deteriorate with age and could present a liability to the safe and reliable operation of the new radio.

Antennas

The COMM antenna should be located perpendicular to the horizontal plane of the aircraft and at least 6 feet from DME and transponder antennas and 3 feet from ADF and other COMM antennas. The antenna should be kept clear of obstructions or projections such as the vertical stabilizer, landing gear doors, engines, and propellers. In some systems with dual COMMs, one antenna is mounted on the top of the fuselage, and the other, an angled wire antenna that runs vertically for about 10 inches then bends to run horizontally, is mounted on the belly. Other installations do fine with both COMM antennas mounted on top of the fuselage, provided there is enough room for the proper spacing. If one antenna is mounted on the belly, the landing gear doors must be bonded with a minimum of two braided straps at opposite ends of the doors to prevent blanking of transmission or reception while the gear doors are open and the airplane is on the ground.

Because some aluminum aircraft skins are too thin to support an antenna, your installer might have to add a reinforcement plate or doubler to the area where the antenna will be mounted to prevent cracking or tearing of the skin. If the skin is composite, a ground plane measuring at least 18 square inches will be required. If the composite skin is thicker than 90 thousandths (0.090) of an inch, the ground plane will have to be installed externally on the skin and secured with fasteners or a layer of Fiberglas. Otherwise, a dielectric effect could result, reducing antenna performance. See Chapter 7 for more information on composite antenna installations.

COMM Radio Problems

Here is a useful flight-test routine to establish the performance of your communications radios:

1. First, review the manuals for your radios to make sure you're completely familiar with how they operate and what kind of performance to expect from them.
2. Before takeoff, operate all flight controls through their full range of travel to assure there is no interference in your radios. Ask for a radio check on both COMM 1 and COMM 2 while constantly listening for static, breakup, and blanketing effects of reception and transmission.
3. Turn on rotating beacons, operate flaps, cowl flaps, and other flight controls while operating the radios.
4. Listen for ignition noise and interference from other electrical equipment and the effect communications have on the navigation radios. Confirm that the communications radios are clear of extraneous noise and that your transmissions and receptions on the radio test are loud and clear, without background static.

Look at the difference replacing the Mark 12 with an advanced design MX-300 will do for your panel and to comply with FCC frequency specifications.

If everything checks okay on the ground, take the airplane up for a flight test.

5. After takeoff, try to contact a facility that is 50 nautical miles away. If your transmitter has a high and low power setting, try the radio check on both settings. Try transmitting and receiving while in a 10-degree bank and make sure you don't lose contact while in the turn.
6. Operate the squelch control and determine if it functions properly and doesn't cause any adverse effects.
7. Try the same radio check test at a distance of 30 miles; the facility you are talking to should hear you loud and clear. If not, continue towards the facility until they can hear you loud and clear and note the distance you are from that facility.
8. Ask for a practice approach at a tower-controlled airport. Put the airplane into approach configuration as though you were going to land. You should have the gear and flaps down, all radios set for the appropriate approach, marker beacon audio on, ADF (Automatic Direction Finder) tracking the LOM (Locater Outer Marker) if there is one. During the approach, observe and listen for noise in the audio, abnormal needle fluctuations in the NAV radios, and any other abnormal indications.
9. Note that these tests are much easier and safer if you bring another pilot along to help fly or watch for traffic while you test the radios.

These procedures are designed to evaluate communications quality during turns, approaches, landing configuration, and at a distance from the ground facility. Any loss of radio contact in a particular radial bearing from the ground facility could be caused by poor bonding or unequal torquing of the antenna base to the airframe, a defective antenna, or airframe obstructions (vertical stabilizer, landing gear, flaps, etc.). Confirmation of these problems will help lead you towards a solution—not always an inexpensive solution, but a solution nevertheless.

Troubleshooting Tips

Keep in mind that if you have dual COMMs, audio and microphone selection is most likely selected through a common audio panel. In most such installations, audio is also amplified as it passes through the audio panel. This common unit can help you isolate the cause of a problem.

If your installation has dual COMMs of the same brand and type, one of the easiest ways to isolate COMM problems is to switch radios. That way, you can isolate whether the problem is in the radio or in the airplane. If it's in the radio, then you don't need to bring the airplane to the avionics shop.

Don't try removing any radios unless you've been shown specifically how to do it. Each manufacturer has its own special method of locking radios into the panel, and while it might seem simple, there is usually a trick to doing it right and a special tool. If you are not a licensed airframe mechanic, you are not allowed to remove and install your radios, unless you are doing so under the supervision of a licensed mechanic. Don't forget that FAR Part 43 requires that all maintenance performed be signed off in the logbooks, so make sure the A&P you are working with signs off each time the radios are removed, installed, and switched. This will also help you keep track of problems and their resolution. Unfortunately, many shops ignore the requirement to sign the logbooks when radios are removed and reinstalled for repair, but the requirement is there. It's just a matter of you insisting that it be done in order to maintain the documentation integrity of your airplane.

Frequently you can clear up a transmission or reception problem by removing, cleaning, and reinstalling a radio. You can clean the flat, metal contacts on the back of the radio by gently scrubbing the metal with a clean pencil eraser. Be careful not to remove any metal from the contacts.

The following paragraphs name some symptoms and suggestions for the cause.

1. **Receiver doesn't work.** Check to see if the microphone is keyed. The easiest way to rule this out is to replace the microphone

with a known good unit. Also check that the radio and audio panel are properly seated in their trays. Is this problem occurring on both radios or just one? Don't forget to check whether the circuit breaker has popped out or the fuse has blown (if it uses fuses instead of circuit breakers). There could also be an antenna or coaxial cable problem.

2. **Garbled reception, breaking up.** Check for a misaligned receiver (where the frequency is slightly off tolerance) or poor shielding. Also check for receiver gain set too high, mic gain set too high, or a defective intercom. If you have an intercom, turn it off, or if it can't be shut off, try using the emergency/spare mic jack.

3. **COMM won't transmit.** The mic key button could be defective. Again, try another good mic. Also check contacts on mic jack. Is the jack making proper contact (see Chapter 5)? Check circuit breaker or fuse. Is the radio fully seated in its tray? Also check audio panel seating. Is the audio panel correctly set to transmit on the radio you're using? There could also be an antenna or coaxial cable problem.

4. **Carrier wave only.** Tower says they can hear carrier wave only, but not voice. This could be from a broken wire in the mic. Try a new mic, and if that works, examine the old mic to see if it's just a broken wire that can be easily fixed (see Chapter 5).

5. **Intermittent transmission.** Try a backup mic. It could also be a defective relay in the COMM radio or antenna problems.

6. **Transmission poor.** Squawked as weak, garbled, breaking up, or noisy. Try backup mic. Also make sure your lips are close enough to the mic. If it's not the mic, it could be a misaligned transmitter or faulty antenna system or coaxial cable problem.

7. **Headsets are not working.** Try using the copilot's jacks or another headset to see if the problem is in the airplane or the headset.

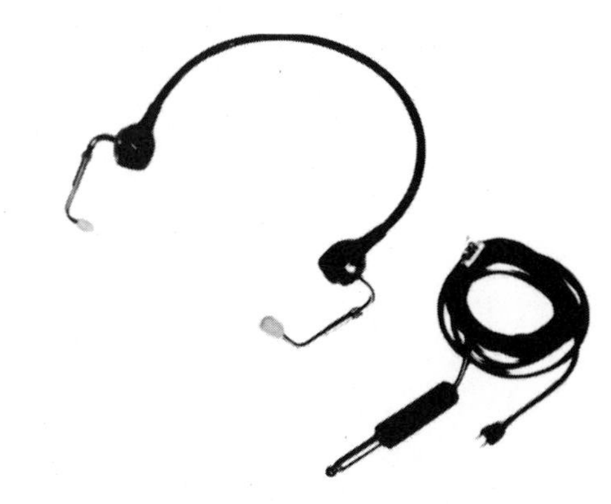

Telex headset.

8. **Speaker inoperative.** Try the headset to make sure the audio system is working. (Don't forget to set the audio panel correctly to listen on headphones.) If system works with headsets, the problem could be a defective speaker, wiring, or amplifier.

COMM and NAV/COMM Manufacturers

The following lists several recommended communications and navigation/communications radios.

Aire-Sciences

Becker Avionics

Collins Avionics

Icom

ICS (available from Wag Aero)

McCoy Avionics (replacement front end for older King radios)

Narco Avionics

S-TEC

Sigma-Tek

Radio Systems Technology (available as kit or factory built)

TKM Michel (slide-in replacements for older King, Narco, and ARC radios)

Val Avionics

Handheld Transceiver Manufacturers

These companies offer handheld transceivers. See Sources for addresses.

Bendix/King

Communications Specialists

Icommunications

McCoy Avionics

Narco Avionics

Sporty's Pilot Shop

CHAPTER 9

Navigation Systems

The VOR system is primarily designed to receive signals that will feed information to drive a needle or other form of steering display, referred to as a course deviation indicator. These systems consist of the receiver, indicator, antenna system, and in the case of remote black boxes, frequency-channeling control heads. All of the systems drive an indicator, whether it is a separate OMNI head in your instrument panel or an LCD display in the panel-mounted radio, like Bendix/King's new KX125 NAV/COMM.

NAV receivers operate on frequencies between 108.00 to 117.95 MHz. VOR frequencies are the even hundreths (110.30, for example) and ILS uses the odds (109.35, for example). This is important to know because ILS frequencies will not work if the frequency common (ground) is missing, while the VOR will work fine. The opposite is true if the ILS common circuit is always grounded; then the VOR frequencies won't work. In other words, to ensure that the VOR is switched off when you are using the ILS frequencies, the ILS common wire is grounded when you tune in an ILS frequency. When you tune in a VOR frequency, the ILS common is ungrounded so only the VOR frequencies will work.

Installation Tips

As with any radio installation, proper assembly of the wiring harness and use of high-quality wiring is a critical step. If this is a retrofit, consider replacing the old wiring and circuit breakers at the same time to prevent reliability problems.

Standard RG-58 coaxial cable is usually used for the antenna system, but when electromagnetic interference is a potential problem, double-shielded cable might be necessary. Most installations use one NAV antenna and a coupler to tie together the two pieces of coax cable from the individual NAV receivers. If you can afford it and there is enough room for two antennas, it would be best to install two NAV antennas and separate coax cables for each NAV receiver up to the receivers, then split with a diplexer coupler.

NAV Antennas

There are two basic NAV antennas commonly used today: the V-shaped dipole, which looks like two pieces of thick wire mounted on the top of your airplane's vertical fin, and the balanced-loop antenna. The V-shaped dipole tends to be more "blind" to reception of signals to the side or perpendicular to the aircraft's flight path. As the airplane approaches a VOR station, the signal will be stable, but as the airplane passes by the VOR, the NAV CDI needle will become nervous. Because terrain obstructions such as buildings or mountains can twist the normally horizontally polarized VOR signals (see Chapter 7), two signals could result, and the dipole antenna will receive both signals and display this as an error (a jerky or erratic needle). The balanced-loop antenna will reject these unwanted signals, whereas the dipole will not.

Balance loop NAV antennas mounted on vertical stablizer.

The balanced-loop antenna comes in two designs: the towel bar, which mounts on a vertical surface like the vertical stabilizer, or the blade, which looks like a normal knife-shaped blade antenna.

Although RNAV systems are less likely a choice today with the GPS dominating the market, the decision to replace a dipole antenna with a balanced-loop antenna could become more pressing when you are having RNAV installed. RNAV systems require antennas that are capable of receiving signals from all directions, because RNAV works off VORs that are not necessarily in the line of flight. You can try testing a new RNAV with the old antenna by flight-testing the RNAV system using raw VOR data without turning on the RNAV functions and looking for errors while in this mode. If none are detected, the odds are that the system should work okay without installing a balanced-loop antenna, but isn't it worth the time and effort to simply install the best antenna in the first place? The initial higher cost of the antenna and installation will eventually be outweighed by the savings in squawks incurred during the system's lifetime.

Replacing the dipole antenna with a balanced loop will increase NAV range by a considerable amount, as much as 25 percent, even when on the ground. Also, if you fly in icing conditions, the blade or towel bar antenna is more resistant to icing damage and won't break off as easily as the wire dipole.

Troubleshooting VOR Problems

If your airplane has dual VOR receivers, take advantage of having both by comparing one to the other during troubleshooting. You can switch receivers and/or indicators to isolate the problem as well.

Just as with communications systems or ADF, initial troubleshooting begins with listening to the audio, in this case the VOR or ILS or localizer identifier. The audio you hear is carried on the same signal that drives the CDI. Therefore, if the CDI isn't working and you can hear the station identifier, the problem is either in the converter, interconnect wiring, or the indicator.

Select a station with a strong audio signal to confirm the frequency you selected. Everything needed for the converter to drive the left/right needle is converted from that signal. Listen carefully for noise. Any noise detected within that audio will tend to override or drown out the signal, in essence producing a relatively weak signal. What you'll see on the indicator is erratic needles. If the audio is loud and clear but the indicator doesn't work at all, the problem is most likely within the converter, which is located either in the receiver or inside the indicator. Obviously, a qualified shop will have to perform repairs to fix this problem.

If the audio is noisy in both receivers, the problem can be isolated to a common system: the antennas. The problem can be anything from a defective antenna to poor RF bonding to broken or kinked coaxial cables. If the VOR is working but the localizer isn't, it's possible that the common wire for the ILS frequencies has broken, which would prevent the localizer from channeling. This wire should be grounded when the LOC (localizer) frequency is selected, but should not read to ground when a VOR frequency is selected.

When testing your VOR system, if you have RNAV, make sure it's set to allow normal VOR operation so you don't try to troubleshoot something that is simply the result of a switch setting.

One area that could cause a problem is the NAV coaxial cables. Periodically, technicians or mechanics accidentally switch the NAV and communications cables get switched during maintenance. Although the COMM and NAV systems can still operate, they won't be as efficient and could cause you to spend money trying to track down the problem elsewhere.

Properly installing radios in their racks is a primary concern when beginning troubleshooting. Many times radios are removed and incorrectly reinstalled so that the rear connector is not seating properly with the rack's rear mating connector. Another possibility is that during original installation, the radio rack was located too far back from the front of the instrument panel, not allowing the connectors to make solid contact. You can check this by measuring from the back of the radio bezel (the radio's faceplate) to the rear of the radio.

Take another measurement from the front of the instrument panel to the connector in the back of the radio rack. Compare the two measurements. Do they indicate there is a compatible relationship between the radio plug and its mating connector? The difference doesn't have to be much to break contact or cause intermittent contact, especially on ribbon connectors or "F" style contacts.

F-style contacts are a terminal created from flat metal stock and formed to create the contact area and the crimp section. The crimp section is left open, forming a trough or "U" to accommodate the conductor. A side view gives the appearance of the letter "E." The conductor crimp is the short portion of the "F," while the longer section or insulation crimp is the upper part of the "E." These F-style contacts are used in many avionics installations, especially panel-mounted radios. A small, protruding tab or finger acts to retain the terminal in the connector block. After a lot of use, this tab wears out the plastic housing and the tab slips loose, releasing the contact. Other failure points are the spring-loading of the contact against the mating connector. Eventually, after repeated use, the spring-loading tension deteriorates and must be readjusted, or the contact must be replaced. Another failure point is where the conductor is crimped. If the wrong crimping tool was used, it may have over-crimped the conductor and virtually cut through the strands to the point the contact will eventually fail.

Antenna Troubleshooting

Troubleshooting navigation antenna system problems in the aircraft is rarely simple. In fact, it is usually a difficult task because of the upholstery, unknown routing, equipment removal, and all the possible failure locations. The most common antennas are the "V," "Whisker"-style, and the balanced loop. Both are found on the vertical stabilizer, except on the "V" tail Beech Bonanza or helicopters. In these installations, the balance loop antennas are directly opposite, on both sides or around the end of the tail. (Balanced loop can have a separation of up to 30 inches using standard RG-58 coaxial cable; if greater spacing is needed, a special cable is required.) A balun transformer provides impedance matching for the "V" design while the balance loop, which is found on higher performance aircraft, uses a matching coupler (Diplexer) that combines and blends the signals from both blades while maintaining signal isolation between the antennas. The combined signal travels along a single coaxial to split using another coupler near the NAV receivers, which connects to both the #1 and #2 navigation receivers. If a Glideslope is to be connected to the NAV antenna, the coupler near the NAV receivers would be a triplexer that would have two NAV outputs and one G/S coaxial output.

The drawback to such an installation is there is one single weak link and that is the coaxial cable running from the antenna coupler to the receivers. If it develops problems, three systems fail. The solution? Run two coaxial cables directly from the antennas to a coupler or triplexer near the radios and then split to each radio. Only one coupler is used, and there is cable redundancy.

VOR signals tend to distort because of buildings, mountains, and temperature waves, and during transmission, thus causing the common and irritating needle oscillation. To confirm that the NAV system is not the problem, a process of elimination (troubleshooting) must be conducted. Troubleshooting requires the technician connect a test set, either radiated or direct. If direct, it must be capable of injecting at least one watt into the coaxial cable. Test sets available are the Michels 1500 ($4,995) or the IFR 401 ($9,995). Both of these prices are unlikely to change for the next five years. This is especially true for the Michels 1500.

Frequencies to check are 108, 113, and 118 MHz. A Bird 43 or equivalent (with a 1-watt loading slug) should be connected directly to the coaxial cable that connects to the radio (coaxial cable is disconnected from the radio). This places the wattmeter in-line between the antenna and the receivers. A VSWR reading of 5:1 or less is acceptable (if 3:1, start looking for coaxial cable or antenna failure). Should the balanced loop antennas have dual coaxial outputs, the one not under test should be connected to a 50-ohm load to simulate the antenna, Reverse the procedure by loading the #1 when checking the #2 coaxial cable. Simply put, load the antenna end of the coaxial cable and inject the test signal at the other end. If a two cable system, disconnect both the #1 and #2 coaxial cables from the receivers and at the antennas. The test is then the same as a one-cable system.

Ineffective ground planes can also cause needles to be overactive. To examine for this, inspect for poor antenna bonding (obvious signs are corrosion, sealer missing), bonding straps missing (or never installed), or frayed straps between the vertical stabilizer and the rudder. Also, check the access panel bonding on the vertical stabilizer and removal tail cone.

If you have two NAV receivers and both are displaying similar indications, such as weak or erratic reception, it is likely that there is an antenna system (coaxial cable, couplers, connectors, or antennas) problem or that there is a bonding problem at the antennas or on

Bird wattmeter.

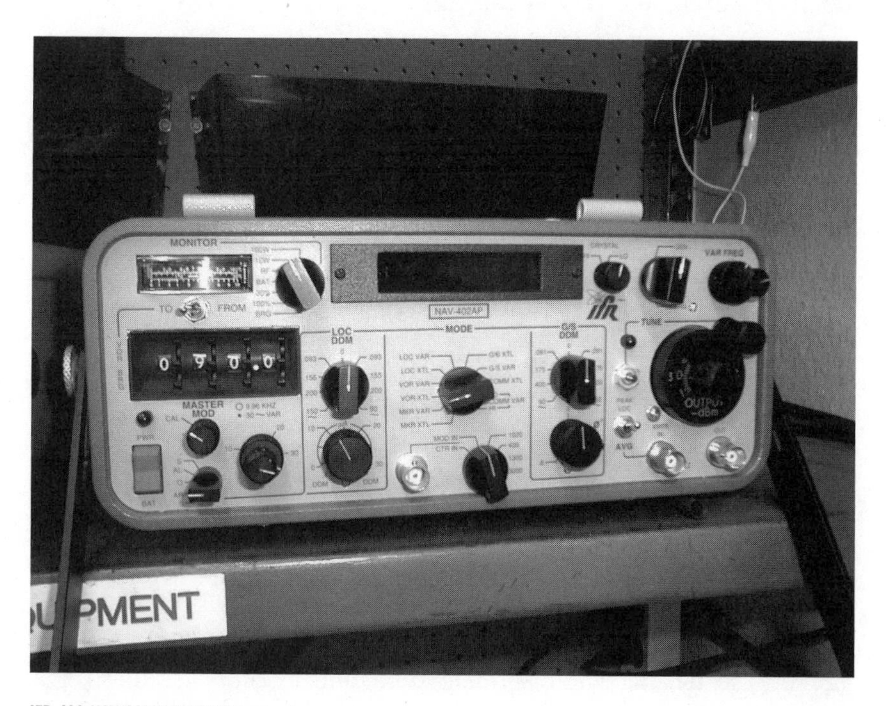

IFR 400 NAV/COMM/MKR/G/S test set.

control surfaces. To test this, the avionics shop will test the system without taking the airplane apart by using a test set and radiating a test signal. They'll situate the test antenna about 12 inches from the NAV antenna on the right side of the vertical stabilizer and radiate a 60- to 70-dB signal, slowly decreasing the dB level until the NAV flag pulls and the CDI (Course Deviation Indicator) needle swings. They'll do the same for the left side of the antenna and compare the readings. With a balanced loop antenna system, found on most light-aircraft and all heavy iron aircraft NAV systems, the readings where the flag pulls and swings the needle should be virtually the same. Any variation of more than 4 to 6 dB indicates a problem with the antenna. In this case, you'd want to change the entire antenna, including the coupler and the short cables between the antenna and coupler.

Here is a simple ground check you can perform yourself. With the NAV radios on and tuned to a nearby station (obviously you need to be at an airport where you can tune in a local VOR to do this test), have a partner take a long broom handle and tap gently on the antenna elements. Watch the CDIs and listen to the audio for any increase in noise or needle movement. This could indicate a defective antenna or coaxial cable connections. Move the control surfaces (rudder, flaps, trim, or ailerons) while watching the CDIs and listening to the NAV audio. You shouldn't see any needle movement or hear any noise while the control surfaces are moving. If there is needle movement or noise in the audio, the surfaces must be bonded to structure in at least two points on each surface. Oh yes, there is the potential of poorly bonded access plates or covers. They can also cause similar problems and this can be isolated tapping gently on the cover with a rubber mallet or vibrate with a sander covered with a felt pad. Of course, there is the possibility of precipitation static producing noise on the surfaces or covers and that is much more difficult to locate. The only solution is to simply start bonding them one at a time until the offending location is detected. Sure, you can use a bonding meter, but that also takes time, and it is just as quick to remove, bond, and reinstall than to add another step. The decision is up to the respective shop or the aircraft owner/pilot.

The easiest way to eliminate or confirm a coaxial cable is the problem is to bypass the existing cable with a length of known, good cable, then test the system using the existing antennas. If this test shows there is a definite coax problem, before replacing the entire length of coax in the airplane, check for bad bulkhead fittings, where the coax is split and connected with connectors to go through a bulkhead. It would be much easier to replace a bad connector or two than to replace the entire length of coaxial cable. If you're lucky, you might find that during manufacture, extra lengths of coax were laid in the airplane, and you might be able to use one of these pieces of coax, saving you a lot of time and effort. If necessary, the cable can be ran on the exterior of the aircraft and secured with high-speed tape (aluminum duct tape), leaving no edges to lift.

VOR Ground Check

If you are a pilot flying IFR, you must have logged a VOR test within the past thirty days. There are two ways to do the test on the ground. One is to use a FAA VOR test facility (VOT) or a radiated test signal from your avionics shop's equipment. Note that if having the check done by your avionics shop, the test must be logged by the shop itself because it's their equipment that is providing the signal. The VOT is like a VOR that broadcasts only one radial. Perform the check by tuning in the VOT frequency and centering the CDI. With a "from" flag showing, the OBS (omni bearing selector) should read 0 degrees. With a "to" flag showing, the OBS should read 180 degrees. At the same time, you should be able to receive the audio from the VOT facility. Maximum error on this test is 4 degrees.

The other VOR ground check method is a VOR checkpoint on the ground if your airport has one designated. If you have dual NAVs installed in your airplane, you can check one against the other using the VOR ground check point. The maximum permissible error between the two NAV's is 4 degrees.

NAV Flight Test

You can perform the VOR test in the air using certified airborne checkpoints; the maximum error being 6 degrees.

Here is another useful test: Fly over a landmark at a right angle (90 degrees) to a VOR within good reception range. Center the needle when you're exactly 90 degrees to the VOR then note your position relative to a good landmark directly below the plane. Turn the airplane around and, leaving the NAV radio alone, check the needle when you fly over the same landmark in the opposite direction but still 90 degrees to the VOR. The needle should still be centered. If not, this might indicate a NAV antenna that's not sensitive to all azimuths. Before replacing the antenna, try rebonding it and equally torque its fasteners; then perform the test again.

If your antenna is the inexpensive V-shaped rabbit-ear dipole, you might want to consider upgrading to the more efficient and reliable towel-bar or blade-type balanced loop antenna. The balanced-loop antenna has a broader directional reception. On the other hand, the V-shaped dipole antenna is most effective in the open section of the V, which is directly behind the airplane (unless installed with the V facing forward).

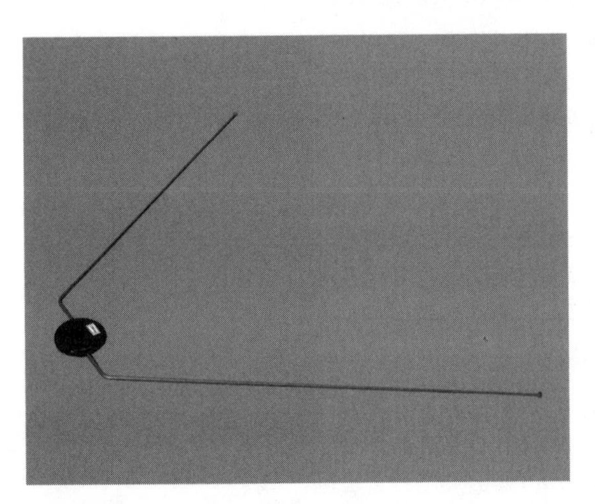

V-style NAV antenna. Not the best choice; reception is not equal in all directions.

An important note: While your NAV radios might be within limits on the VOR test, don't assume that means they are working perfectly. As you fly toward the range limit of the NAV radios, any error will be magnified. At the maximum reception range for a VOR, for instance, the error could be unacceptably high because of the weaker signals at that distance. If you rely heavily on your NAV radios for IFR flying, consider having them recalibrated by your avionics shop once a year to prevent the loss of accuracy as signal strength diminishes. Locations and frequencies of certified VOTs and ground and airborne test points are listed in the FAA's Airport Facility Directory.

Area Navigation

This discussion of RNAV is limited because it is becoming somewhat of a dinosaur with the advent of sophisticated new loran and GPS sets. RNAV is still useful because its use is fully integrated into the current ATC system, but its high price makes it nearly as expensive as an IFR loran, and the loran can do a lot more for your money. GPS IFR approaches look like a certainty in the not-too-distant future, so in comparing RNAV to GPS, both in terms of price and features, GPS wins hands down.

What follows is a ramp checklist, courtesy of Bendix/King, for testing RNAV system accuracy. This test allows you to perform a simple ground test, eliminating the need to have a shop drag out their expensive equipment. Most shops use this test anyway.

Before beginning this test, make sure strong DME and VOR signals are available and that they both work with the RNAV set to VOR mode (both DME and VOR signals are necessary for operation of RNAV systems).

Test 1

1. Select VOR mode on RNAV.
2. Record radial to VOR station by centering CDI with "to" indication.
3. Set the waypoint distance equal to the DME distance indicated to the VOR.
4. Select RNAV en route mode.
5. Rotate the OBS until the CDI centers with a "to" flag.
6. RNAV distance should read 1.41 times DME distance (plus or minus 5 nautical miles) and indicated course should be 45

degrees (plus or minus 2 degrees) greater than the VOR radial recorded in step 2.

Test 2

1. Select VOR mode on RNAV.
2. Record radial to VOR station by centering CDI with "to" indication.
3. Program a waypoint with a radial 120 degrees greater than the indicated VOR radial.
4. Program the waypoint distance equal to the indicated DME distance.
5. Select RNAV enroute mode.
6. Rotate the OBS until the CDI centers with a "to" flag.
7. RNAV distance should equal DME distance (plus or minus five nautical miles) and indicated course should be 60 degrees greater than the VOR radial recorded in step 2.

Glideslope Receivers

Because glideslope receivers are usually built into the NAV box, failure of the glideslope might be a fault of the NAV receiver. Obviously you would notice if the NAV failed and make the correct assumption that the glideslope is also inoperative before trying to use it on an approach. The glideslope is fairly simple in design and layout, so if it has failed, it shouldn't be that difficult for your avionics shop to fix. Typically, more problems are caused by shielding integrity, bonding, and proper installation than simple wiring failure.

Other transmitters can easily affect the sensitive glideslope receiver, so make sure all antenna connections are secure and that there is sufficient distance between antennas (2 to 3 feet minimum).

Some systems interface the NAV and glideslope to the NAV antenna with a coupler. If the NAV is working but the glideslope isn't, check the coupler to the glideslope receiver. It's possible that the coaxial cable was disconnected during maintenance, say when the nose cone was removed, or the center conductor has worked its way out of the coax connector (poor installation or handling).

On single-engine airplanes where the factory or the installer has placed the glideslope antenna just inside the windshield, you might see a problem with the glideslope needle wavering slightly up and down from propeller modulation. Except for confirming that the receiver is fine, the coaxial cable is intact, and the antenna is in good shape, there isn't much you can do about prop modulation. You might try changing the coaxial cable or the antenna. A better solution would be to install an external glideslope antenna away from the propeller's effects, or tie the glideslope receiver to the NAV antenna with a coupler.

Michels Model 1500 can provide the test signal to assure proper glideslope operation.

NAV Receiver Manufacturers

These navigation receiver manufacturer's addresses can be found in the Sources section at the end of this book.

- Aire-Sciences
- Becker Avionics
- Bendix/King
- Collins Avionics
- ICS Plus (available from Wag Aero)
- Narco Avionics
- Radio Systems Technology (available as kit or factory built)
- S-TEC
- Sigma-Tek
- TKM Michel

CHAPTER 10

Marker Beacons

Marker beacons are one of the simplest pieces of avionics equipment in your airplane. One of three lights and a distinctive audio tone notify you that you're flying over a marker transmitter. The marker display is usually included in the audio panel, although you can buy it separately. The entire system consists of an antenna, coaxial cable, audio circuits, a receiver, and a switch or two.

The white marker light is for the inner or airway marker; you'll rarely see this one light up. It is activated by a 3000-Hz tone from the transmitter. The middle marker emits a 1300-Hz tone and lights the amber light on the panel and is found about 3,500 feet from the end of the runway. The outer marker is a 400-Hz tone that lights the blue lamp on your panel at the point where you normally intersect the glideslope and begin descending, 4 to 7 miles from the end of the runway.

Marker systems are very sensitive to long cable runs or poor antenna location, especially if the antenna is being shadowed by other antennas or obstructions.

Marker beacon antenna mounted on the aircraft belly. Some are mounted inside the radome. In all cases, it is best to locate as far forward as possible.

Installation Tips

The antenna should be located near the front of the aircraft, with the coaxial cable as short as possible to reduce signal loss to the receiver. Although engineers try to have the antenna as close to the receiver as possible, the design of the antenna and the fact that the marker transmitters are directly underneath the airplane when the signals are received dictate that the antenna be installed on the aircraft's belly. It should not be mounted any further aft than where the belly sweeps upwards toward the tailcone.

The antenna is usually a long blade or boat-style, but you'll still see the older-style (cheaper), thin, towel-rack type rod with a sliding adjustment mounted to the bottom of many Cessnas.

Marker Problems

Failures associated with the marker beacon system are usually related to broken wires, failed antenna systems, burned-out bulbs, or receiver degradation caused by age and vibration. You've probably noticed that the blue light is difficult to see, even under normal lighting conditions, but especially when the sun is shining on the panel. To get used to what the light looks like, shadow the light with your hand while you check it with the test switch. You might be able to get a bench tech to change the filter on the light to change its perceived brilliance.

Marker beacon receivers have simple antenna and coaxial cable requirements. The most common pilot squawk is that the outer marker (OM) is not received or they hear only the audio. This most likely occurs because of receiver degradation. When the aircraft is above the OM transmitter, there is a lot more distance between it and the plane. However, as the plane continues to descend on the glideslope while approaching the middle marker (MM) and the inner marker (IM), the plane is also quickly moving away from the OM outer marker, maintaining the original ineffectual signal distance. As the plane is landing it is quite far from the ground antenna and it doesn't get any closer as it approaches the other two ground marker antennas. At this point, the receiver must be at its most efficient receptive capability. If it is not, there may be audio, but there won't be an OM light. Simply stated, the reason for the light not illuminating is the marker beacon may not be operating at maximum efficiency. As older marker beacon receivers degrade due to age their reception efficiency decreases.

Don't forget to set the sensitivity switch to the "high" position when checking marker beacons in the air and during instrument approaches. You wouldn't want to squawk a market reception problem that is caused by a simple incorrect positioning of the switches. By the same token, whenever flying VFR and especially if your airplane is flown by VFR pilots who don't know how to use the marker beacons, make sure the marker audio is switched off. One accident happened to a student pilot who panicked when the marker started flashing and beeping at him as he flew over the marker transmitter. The student thought something was seriously wrong with the airplane and performed a not-so-flawless emergency landing.

Vibration eventually causes the center contact on the bulb to wear out where it touches the spring-loaded contact in the socket. Replace the bulb when this happens. If a bulb is not available, and you are solder-qualified, resurface the center contact with new solder. If you use too much heat, the bulb can easily be damaged. This is not an recommended method but simply an alternative when a replacement bulb is not available. It should be logged as defective and be replaced when the bulb is in stock.

When the antenna starts to deteriorate, the first light to lose its brilliancy is the blue outer marker. Cause of the problem could be a bad antenna or simply poor bonding. Weak reception might be nothing more than factory tolerances having slipped. First have the avionics shop bench check the receiver, then test the system in the airplane to isolate the problem. If the tolerances are not up to spec, the shop will adjust them as required.

During testing, it's nearly impossible to activate the marker receiver with the test equipment antenna inside the cockpit. Except for direct coupling tests, the technician will usually have to place the test antenna within 12 inches of the marker antenna on the belly. If the tech has to place the test antenna closer than 12 inches to activate the marker beacon receiver, it could indicate that the receiver's switches are set to "low" sensitivity, a defective or weak antenna, shorted coaxial cable, an open circuit in the receiver, or receiver needing calibration.

If the receiver tests okay, have the antenna rebonded or do it yourself (see Chapter 7). If antenna rebonding doesn't help, replace the antenna and coaxial cable. In most cases of marker problems, the antenna was the problem, especially when it was the old-style towel-rack antenna previously mentioned.

The author has found that the sled-type antenna, such as the Comant Cl-102 and other similar types, are the most common causes of weak markers. The Cessna rod-type antenna, which must be tuned when installed or suspected of being weak, has proven to be very reliable.

Once the plane is brought into the shop to resolve the marker squawk, the technician will conduct a ramp check to verify the complaint. First, the technician will gain access to the antenna and remove the coax from the antenna, connecting it to a suitable MKR generator. A good rule of thumb is to use the sensitivity specifications for the marker receiver in the Bendix-King KMA-20 audio panel, which is 500 microvolts. This should activate the light and provide good audio in the high-sense mode and deliver 2000 microvolts for a light and audio in the low-sense mode. Allowing 3 dB of loss for the cable, it should not take more than about 700 microvolts hard for proper high-sense operation or about 2800 microvolts for proper low-sense operation at the input to the RF coaxial cable. This go/no-go test effectively breaks the system in half and points you in the direction of the fault.

Michel Model 2200 can provide a simple but accurate method of testing the marker system.

Marker Beacon Receiver Manufacturers

These companies manufacture marker beacon receivers. See their addresses in the Sources section of this book.

Bendix/King	Narco Avionics
Collins Avionics	Sigma-Tek

CHAPTER 11

Automatic Direction Finder (ADF)

The automatic direction finder (ADF) might seem to be an anachronism in these days of ultra-precise GPS and reasonably precise loran systems, but ADF is still a useful, simple, and widely used navigation system. Fancy NAV systems get all the glory, but ADF is frequently the NAV system of last resort for Alaska bush pilots and over-water ferry pilots. ADF is still the NAV system of last resort, with the exception of GPS, in the former Soviet Union. In fact, in 1999 several U.S. government agencies developed a fully operational system using multi-frequency shift keyed (MSK) modulation at low-frequency signals of 285 to 325 kHz. This method reduces noise to a point that the system is accurate for ADF to within 1 meter near the reference stations. One drawback is that it degrades 1 meter for every 150 kilometers.

The frequencies transmitted by the NDBs (nondirectional beacons) that drive your ADF needle are in the low-frequency range of the radio spectrum, 190 to 1750 kHz, which happens to include the same frequencies used by commercial AM broadcast stations. The ADF will not only receive the commercial stations' audio signal but can hone in on their powerful signals as well. This provides the pilot with a backup means of verifying position or course, if the frequency of the broadcast station and its antenna location are known.

KR-87 ADF receiver.

Be careful about using commercial stations at night, because the higher frequencies they use tend to bounce off the atmosphere and not hug the ground. The signals received by your ADF should be ground waves that hug the ground, which lower frequencies tend to do, even at night. Sky waves that bounce off the atmosphere are not accurate because they don't always bounce straight.

Because the ADF can always get some audio signal, many technicians find it useful as a troubleshooting aid for checking audio quality and output. This way, the tech doesn't have to constantly call the tower or someone in the shop for a test frequency for radio checks. Technicians are often heard saying, "The ADF isn't working; this is going to be a long day!"

Another use for the ADF is finding the LOM (locater outer marker) during Instrument Flight Regulation (IFR) approaches. You get double confirmation when you fly over: a flashing blue light and audio tone from the marker beacon, and an abrupt reversal of the ADF needle.

Many pilots have found the ADF useful in determining the location of thunderstorms. Using an off-station frequency where the needle simply goes in circles because there is no human-generated signal, you can sometimes watch the needle point in the direction of the huge electrical discharges produced by lightning. If you have the ADF audio turned on, you can hear these discharges as well; they sound like loud scratchy waves.

ADFs are susceptible to interference from varying weather conditions, time of day, and even geographic location, so keep these effects in mind when troubleshooting ADF problems.

Front-end for King ADF

McCoy Avionics makes a digital-display, multiple-memory front-end replacement for Bendix/King's KR85 ADE. The MAC 1850 front end includes storage for ten ADF frequencies, plus a countdown/up timer that has a vocal callout feature. This allows you to listen to the timer countdown on an IFR approach while you concentrate on flying the airplane, and other useful features.

When considering this kind of upgrade to your ADF, don't forget that the new front end is only going to make the radio as good as the original radio's guts (electronics). If the radio isn't in reasonably good shape already, then adding this front end won't make it any better.

Installation Tips

Ask your avionics shop if they will include in your price any adjustments that need to be made to ensure accuracy of your ADF system after it's installed. This applies particularly to quadrantal error described later in this chapter.

ADFs are particularly sensitive to extraneous noise, and correct routing of the ADF's wiring is critical to avoid noise problems. Normally standard RG-58 coaxial cable is satisfactory, but double or triple shielded coax made by solid manufacturers such as Belden®, Times, or Essex® is recommended. ADF harness wiring should not be routed parallel to or in close proximity to distance measuring equipment (DME) wiring, transponder wiring, or any power-carrying wires, especially those from the alternator. All necessary wire crossover should be executed at right angles to reduce the possible induction of unwanted noise.

ADF Antennas

ADF antennas have evolved considerably, from the old loop antenna that actually rotated, to the modern combined loop and sense antenna in one streamlined unit. If you're replacing an ADF and your

ADF sense/loop antenna removed from plane with connector pins exposed.

airplane has the less-than-modern wire-sense antenna that goes from the tail to the top of the fuselage and the loop antenna on the belly, now is the time to upgrade to the combined loop and sense antenna. It works a lot better and is much easier to maintain.

The ADF antenna should be mounted 4 to 6 feet from DME antennas and at least 3 feet from COMM antennas. It should be kept clear of obstructions or projections such as the vertical stabilizer, landing gear doors, engines, and propellers. Any openings such as access panels will break the smooth ground plane required by the antenna, so keep the antenna as far away as possible from access panels. If this isn't possible, make sure the access panels are well bonded (electrically) to the airframe.

Installing the combined loop/sense antenna on the belly or top of the fuselage won't make any difference when the airplane is flying. The belly location, however, will reduce signal quality on the ground. If installing the antenna on a low-wing airplane, the antenna should be mounted aft of the wing. Some loop/sense antennas, like Bendix/King's, have drain holes. If mounting the antenna on top of the fuselage, plug the drain hole with silicone sealant.

As usual, if the mounting point isn't strong enough to accept an antenna by itself, the installer might have to add an aluminum doubler to the area before mounting the antenna. The doubler must be RF-bonded to the airframe.

Troubleshooting ADF Problems

The best way to operationally check an ADF receiver is to select a strong commercial frequency and check for correct needle pointing. If the needle points to the station correctly, select the NDB (nondirectional beacon) frequency on which the ADF was squawked as giving poor performance. Assuming the higher-strength commercial station works and the lower power NDB doesn't, the problem might be a defective antenna system or defective receiver. Have the receiver bench-checked before tearing into the antenna system.

A short-circuited antenna, failure of the RF section of the receiver, or defective wiring interconnect can cause weak reception. If you suspect any of these problems, have the receiver bench-checked. Of course, other problems such as obstructions between the aircraft and the station, onboard noise caused by other equipment, and ambient weather conditions can also limit ADF reception. Being a low-frequency signal, reception of NDB signals is much more sensitive to atmospheric conditions than the higher-frequency VHF signals used for communications and VOR navigation.

If the pointer doesn't point to the station at all, the problem might be wiring or equipment failure. Either problem can cause the same symptom, requiring isolation of the problem by point-to-point checking of wiring and bench-checking of the ADF unit. Don't ignore the possibility that the NDB might have been out of commission for the time during which the ADF didn't work on that station.

The alternator can be a big source of ADF interference and can wipe out several frequencies or even just the lower band of frequencies. If the ADF displays signs of failure with the engine running but works fine with the engine off, try shutting off the alternator switch while the engine is running and see if that makes a difference in the reception. To perform this test, shut off all radios and electrical equipment before shutting off the alternator. Then shut off the alternator, if you have a switch to do so. If not, you might have to pull the alternator field circuit breaker to take it off-line. Turn the ADF back on and note whether the ADF reception improves with the alternator off. Before you turn the alternator on again, make sure the ADF and all other electrical equipment is off first. Turning the alternator on and off can create an electrical surge that could damage sensitive equipment, especially avionics. Recommend that the standard filter on the alternator be removed and a new one from Lonestar be installed (Part #LS03-01004). The new capacitor has double the capacitance of the original capacitor and therefore much better filtering capability of alternator whine.

Noise from other sources such as the DME and transponder can affect ADF operation. Some DMEs, such as the one that is part of older versions of King's KNS80 RNAV, lack a modification that prevents noise being injected into the ADF receiver. You can check for interference from the DME and transponder by checking ADF operation with DME and transponder on and off, or both on and both off. If you suspect either one is causing a problem, talk it over with your avionics tech and ask if there are any modifications that should be done to the offending radio to prevent this problem. If the ADF needle rotates continuously and doesn't point at the station, check for breaks in the sense antenna (the long wire antenna from the vertical stabilizer to the top of the fuselage, on airplanes where the sense and loop antenna aren't combined in one unit).

Check for missing audio by setting the switch on the ADF box to the ANT (or antenna) position. Missing audio may be traced to the loop antenna or the receiver. Being on the airplane's belly, the loop antenna is susceptible to oil and water contamination, especially when the airplane is being washed, causing water to find its way into the mating connector. Narco antennas have a built-in cup that will hold quite a bit of water and still allow the antenna to work for a while, but not very well. Cleaning the water off and reassembling the connector might help, but possibly only temporarily because damage could already have been done. Narco provides a sealer kit to prevent moisture entry, so if this isn't installed on your Narco antenna, you might want to have it done. Make sure the antenna is thoroughly dried and cleaned before the modification; you don't want trapped moisture under the sealer.

Troubleshooting ADF problems can be tricky because there are so many factors that can affect the ADF. The pilot has the responsibility to communicate with the technician to help the technician do his or her job efficiently and accurately. Both are a critical part of the troubleshooting link. Let the technician know the time of day, type of weather, altitude, speed, and what other equipment was on when the problem showed up. When the malfunction occurs, try to shut down other pieces of equipment to see if you can isolate the problem to a specific system that is interfering with the ADF.

ADF Loop/Sense Cable Systems

Older automatic directional finders (ADFs) such as the Bendix T-12, Bendix/King KR-85, and the Cessna R-546 had comparative internal electronic circuitry and needed sense and loop antenna cables that were properly matched. When the receiver, indicator, and loop antenna are confirmed as good by bench test, and there is adequate reception of weak signals, the sense and loop antenna cables must be tested. First find what the manufacturer's specifications are for the capacitance, then have the cables measured for correct capacitance, compared against a known good section of cable like that used in the antenna system. This way you compensate for any test equipment variations. Because manufacturers do not specify total capacitance in loop cables and when it is necessary to measure total capacitance you need to measure a known good loop cable and compare it to what is in the plane or better yet, and if practical, substitute the entire cable system.

There is still one coaxial cable that can be a source of problems, and it is located between the receiver and the indicator. Loop signals, from the antenna's two loop windings, are summed in the indicator's goniometer with the resultant signal traveling to the receiver for further processing through this cable. In the KR-85 system, this cable is specified to have only 48 picofarads (pf) of capacitance. Use the method previously described to measure its capacitance.

Visual inspection goes a long way to locate cable failures. Check for nylon tie wraps that have overly compressed the cable (garrote effect), sharp cable bends, connector corrosion, or recent modifications that might relate to the present squawk. One thing that could be happening is the center conductor may have migrated towards the shield, which is not obvious from a visual inspection. If there is a suspicion that this might have happened, substitute with a known good length of cable. The compressed cable changes the capacitance, just as any variation from the cable's normal design would affect the capacitance. Sharp bends and center conductor deflection from cable center cause the same problem.

ADF System Operational Check

Here is a checklist for operationally checking your ADF system. This should be done after a new ADF installation and after any repairs.

1. Check all power and ground connections before turning on power.
2. Turn on power and turn on the ADF.
3. Listen to the ADF audio in both the ANT or antenna position and the ADF position. In the ADF position, the receiver should function as a directional finder, pointing toward a known, good station. ANT or antenna position allows the radio to operate only as an audio receiver. I recommend identifying the station with the switch in the ANT or antenna position first, then switching to ADF mode.
4. After switching to ADF mode, the needle should point directly to the station. During daylight hours, a flight test will have to be conducted to get an accurate test of the system. There are

two types of errors you need to check for: constant and quadrantal. A constant error is the difference between the correct bearing and that which is displayed on the indicator. Fixing a constant error might only require adjustment of the receiver. Antenna adjustment might be possible, but the antenna will have to be removed to make the adjustment. After realigning the ADF, select an NDB station with a known magnetic heading from your position. The ADF needle should point to that heading, indicating a relative bearing of plus or minus 3 degrees.

5. Quadrantal error isn't constant and usually varies in each 90-degree quadrant and depends on the receiver quality and location of the loop antenna. Some receivers allow for fine calibration to remove quadrantal error, and it could take many test flights to make these adjustments.
6. While still in flight, switch between the ADF and ANT positions, leaving the switch in the ADF position. Did the needle move, without sluggishness, directly to the station? Make sure the needle isn't reversing (pointing opposite to where the station is).
7. If a compass card is connected to your ADF system, make sure the card follows within 2 degrees of the ADF needle when you turn.
8. Turn all other avionics off and check if the audio is clear of extraneous noise caused by nonavionics electrical equipment like strobes or electrical actuators, etc.
9. Turn other avionics systems on one at a time and check for noise and interference problems. If a problem is found,

Strobe light.

that system will have to be checked to see why it is suddenly causing an interference problem with the ADF. It could be that another system had a problem all along and it wasn't noticed until the ADF was installed. Any time alternators, generators, strobes, electrical actuators, and control surfaces are replaced, make sure all bonds and filters are reinstalled and are in good shape. (Should the alternator be found as the source of the noise (whine in the audio and spinning needles) don't install a filter in-line of the alternator output. Either replace the capacitor on the alternator or the alternator itself. Of course, it could also be poor grounding of the alternator to aircraft structure.)

10. If a new autopilot was installed recently, check the ADF with the autopilot turned on, engaged, and tracking a nearby VOR or the heading bug. Listen to the ADF audio and watch the needle. Now turn the autopilot off. Was there any change? Did the ADF audio improve or the needle swing to a new heading? If it's a Century autopilot, the autopilot's 5-kHz oscillator (in the autopilot amplifier) might be generating interference through the common bus, the grounding system, or by radiation. For this type of interference problem, it is especially important that the ADF antennas, the receiver, and all grounds adequately bonded to the airframe before attempting any other fixes. Why? Until the autopilot was installed, there wasn't an interference problem. With the new autopilot, however, the ADF is particularly sensitive to signals in the frequency range of noise and the autopilot's 5-kHz generator.

The answer to this problem also might be relocation of components, either the autopilot or the ADF. Or the problem could be solved by relocating harnesses, making sure power lines are shielded, or installing filters in the power lines at the source of the noise, or if impractical, at the receiver of the noise. Another possibility is to lift the autopilot's amplifier off of the ground, eliminating the contact and conduit between the noise generator and the aircraft's chassis.

If nothing else works, try relocating the ADF antenna.

One other slim possibility is that the noise is getting in through wire shielding. This is called a parasitic oscillation or ground loop, and it can occur in cable shielding, even with both ends grounded. To isolate this, lift the shields from ground, one at a time, until the problem goes away.

After reading this, it might appear that this is nothing more than a shotgun approach, trying everything until it finally works. With some problems, it could be that a shotgun approach is the only method that will finally solve the problem, but usually the rebonding, grounding, bus isolation, and antenna relocation can resolve the problem. However, note that these measures should have been done during installation.

ADF Manufacturers

These manufacturers offer ADF receivers. Their addresses are in the Sources section of this book.

Becker Avionics

Bendix/King

Collins Avionics

McCoy Avionics (front-end replacement for Bendix/King KR85)

Narco Avionics

S-TEC

Sigma-Tek

CHAPTER 12

Distance Measuring Equipment (DME)

Although the DME system had been in general aviation aircraft for some time, it is quickly moving out of the limelight. There won't be too much discussed on this particular radio, however, it wouldn't be fair to totally discount the DME; there are still a lot taking up residence in pilots' radio and instrument panels.

Distance measuring equipment (DME) is derived from the military tactical air navigation (TACAN) system. Where TACAN provides both distance and navigation bearing, DME displays the actual slant range distance (within 0.1 percent) from the aircraft to the station. Slant range means that the distance measured is directly from the aircraft to the station, so that if you are flying 5,000 feet above a DME station, your DME will show that you are one mile from that station. Until the advent of GPS, DME has been the only truly accurate

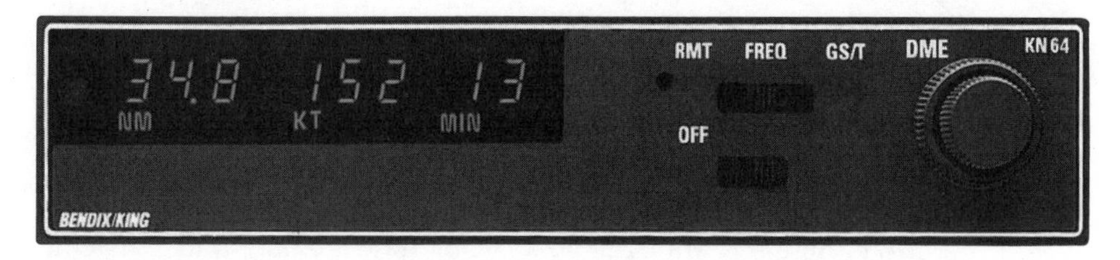

KN-64 DME receiver.

Panel-mounted, remote KDI 572 DME indicator.

distance-measuring type equipment available to light aircraft. With GPS, we may see the end of the DME.

DME uses highly accurate measurement of a UHF (ultra-high frequency) pulse transmission to determine distance to a ground station. The on-board transmitter sends a pulse to the ground station and receives a pulse in return. The aircraft's unit computes the time delay between the two pulses and converts it into a distance, time, or speed display on the DME indicator. Because of the DME's precision, it was used with area navigation (RNAV) systems as well as integrated systems containing loran and GPS sensors.

Installation Tips

Because of the operating frequency range of the DME, its antenna doesn't require the same ground plane requirement as the ADF and VHF. As a precautionary measure, however, it is strongly suggested that the same installation and bonding techniques that would be used for any antenna installation. This has proven to prevent interference from noise generators, the transponder, other avionics, or possible intermittent loss of range caused by precipitation static.

DME is similar to transponders in that they are both pulse systems. The key difference is in the interrogation process. The DME interrogates the ground station and receives a return signal that provides the information used for the display. Transponders are interrogated by radar systems on the ground.

Because DME and transponders operate similarly and on frequencies that are close together, the two systems can potentially interfere with each other. This problem is most prevalent when the two systems' antennas are closer than 3 feet together. If the two systems are interfering with each other, it might be necessary to install a suppression cable between the DME and transponder. The suppression system shuts the DME off for a very short period of time while the transponder is being interrogated to prevent the DME from interfering with transponder operation. With proper antenna placement, installation of the suppression system usually isn't necessary. Many avionics shops will install the suppression system but not hook it up unless testing later shows that the DME is interfering with transponder operation. This approach was applied to many Beech Aircraft planes built in Salina, Kansas, back in the 1970s and early 1980s. Supposedly, the same was done to many Wichita-built planes. You or the technician will have to investigate further to confirm either way. Obviously, this can save you lots of money by allowing the suppression system to be hooked up, if necessary, without ripping up the upholstery to install one if it was left out. It isn't a good idea to hook up the suppression system if not needed because it can reduce the performance of the DME.

DME antennas will resemble the inexpensive lightning rod with a ball, tube, or a stubby, small fin-type blade. Transponders and DMEs can use the same antenna, and both types are installed on the belly of the airplane. The less expensive rod antenna performs fine but is much more susceptible to damage, especially by ham-fisted airplane washers or careless mechanics. The more expensive blade antenna is a better choice for both systems. It will withstand considerable stress, including those doing a sloppy job washing the plane. The DME antenna should be installed 3 to 4 feet from the transponder antenna and at least 3 feet from any ADF or communications antennas.

The older DME units are highly sensitive to low-current availability, so be sure that the proper size wiring is used in the installation. Errors caused by improper wiring may cause only a small change in distance readout, but they will have a greater effect on speed readout.

DME Problems

Important note: Because of the high power output of the DME, potential damage can occur to the DME's receiver/transmitter if the unit is turned on with the antenna coaxial cable disconnected. Although some units are designed to transmit harmlessly with the antenna disconnected, most are not. To avoid an expensive trip to the shop, don't operate the DME with its antenna disconnected.

When experiencing intermittent DME operation, check to see if the antenna is causing a problem by can tuning in a DME station while the plane is on the ground. With the DME tuned in, try gently tapping the DME antenna with a long wood dowel or broom handle. Do not touch the antenna while the system is on. If the rod type antenna has been bent or broken, the distance readout will drop off and start again while you tap the antenna. Believe it or not, I've seen cases where someone, an airplane washer or mechanic, broke the rod-type antenna and simply stuffed the rod back into the antenna base, hoping no one would notice it had been broken. Sometimes these broken antennas would work for a while, but they always had to be replaced.

The tougher blade antennas can experience other problems. Some of the older Narco blade antennas delaminate internally, causing the metal element to vibrate until the fragile connection to the internal connector is broken.

When experiencing DME problems, first try tuning another station. DME stations can handle only a certain amount of interrogators, and if there is one too many, the DME won't be able to lock onto the station. Before spending too much time on troubleshooting a faulty DME, ask the avionics shop to perform a ramp test with their test set. If the system still shows no signs of life, go through the following tests.

1. If it's a ground-speed problem, remove the unit and have it bench-checked.
2. If the DME doesn't function at all, check for adequate voltage and ground.
3. When the DME breaks lock with a station or seems to be weak, check to see if you can receive the audio signal (Morse code ID every 30 seconds). Also, were you flying directly to or from the station? Accuracy is better during a direct, head-on approach. Remember that the DME computes slant range, so accuracy is quite good at a distance, but both distance and ground speed information are affected by how close you are to the transmitter. When you approach the station, the ground speed readout will slow, then rapidly increase as you fly over the station.
4. Reception of a strong audio signal from the station is a positive indication that the antenna system is intact and that the failure is most likely in the receiver/transmitter. This isn't always true, however, because a poorly bonded antenna can cause DME problems. For this reason, it is important that a DME antenna that is mounted on an access panel be well bonded to the panel and the panel bonded to the airframe.
5. A frequent problem squawked is the DME isn't working in the "remote" switch position. This is probably because the pilot didn't have the NAV 1/NAV 2 switch set properly. While the problem could be in the wiring for that switch, the first thing to check is for the switch being correctly set.
6. Sometimes the DME works fine with the engine running but not on battery power alone. If so, the DME's power supply has deteriorated or undersized wiring was installed (or both). If the wiring was the wrong type, the DME may have worked fine initially when it was new, but as it gets older, the need for an adequate supply current becomes greater. The longer the distance between components, the stronger the need for heavier-gauge

wiring and coaxial cable. Don't scrimp on this during the installation.

7. Conduct a time domain reflectometer test on the coaxial cable between the antenna and the TR unit. Your trusty Bird wattmeter won't work to solve cable problems, because the transponder and DME do not produce or receive continuous wave (CW) signals.

Checking the quality of transponder and DME coaxial cables in the aircraft is one of the more challenging jobs. This is because the signals involved are low microwave and at frequencies hard to measure, with high peak power and short pulse duration. This where the TDR (time domain reflectometer) comes in handy, it produces a square wave that travels down the coaxial line looking for reflected energy from damaged cable, sharp bends, etc. The small amount of energy bounced from these disturbances is reflected back to the TDR where the disturbances are recorded and displayed in terms of time. Time is therefore converted into distance, thus providing where the location of the failure. All the information is recorded for maintenance repair or training.

DME Manufacturers

The DME manufacturers' addresses listed below can be found in the Sources section at the end of this book.

Bendix/King

B.F. Goodrich Flight Systems

Collins Avionics

Narco Avionics

S-TEC

Sigma-Tek

CHAPTER 13

LORAN Systems

With the mid-continent gap closed, the only thing that's keeping pilots from installing LORAN (Long-Range Navigation) has been the emergence of the amazingly popular global positioning system (GPS). But this might be premature. There are a lot of LORAN receivers in planes, and pilots weren't happy to hear that the FAA planned to shut down the entire system in the year 2000. With all the improvements, the LORAN navigation system was quite accurate, for the most part. The low frequency of LORAN removes the VOR line-of-sight problem. The entire United States is covered by twenty-three stations that give an average accuracy of less than 1/8 mile, however there are limitations depending on the data base used (bigger is better). As it stands now, the system will remain in operation until the year 2002.

Most LORAN aviation receivers were pre-programmed in the locations of VOR, NDB, Intersections and Airports. The FAA had certified some non-precision approaches for use with LORAN-C, but has discontinued LORAN approaches in favor of GPS. Avionics companies no longer make LORAN receivers, but good used King KR-88 units are available under $900. Most LORAN and GPS manufacturers such as Trimble and Magellan have left the aviation business.

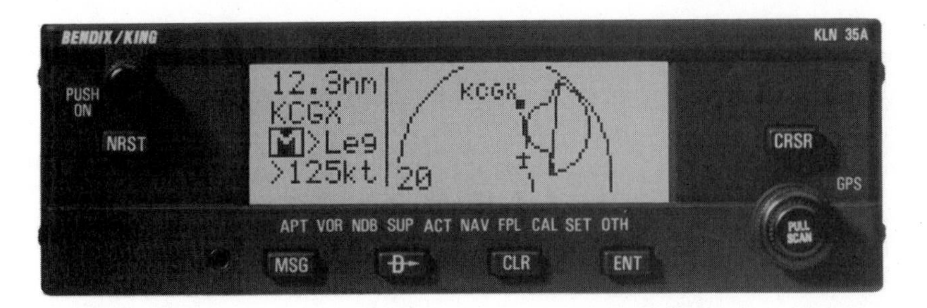

KLN 35A LORAN receiver.

A lot of indecision still exists in the political arena regarding LORANs and GPS. The FAA wishes to wait until the GPS system and the wide-area augmentation system (WAAS) is available to users before determining if LORAN is needed for the aviation system. The LORAN system is presently managed by the U.S. Coast Guard, which (I disagree here—possibly, because we in the industry think of organizations like this as individual entities) would like to turn it over to the FAA.

LORAN transmitter chains are still expanding throughout South and Central America, the Caribbean, Canada, and the Commonwealth of Independent States (former Soviet Union). The continued financial investment in LORAN transmitter sites through much of the world suggests that we can expect to see LORAN around for years to come. Actually, the system was scheduled to be turned off in 2000, but the owners of LORAN receivers insisted that it be kept operational (even at an annual cost of $27 plus million a year). GPS is not ready yet, therefore, the Coast Guard is planning to spend over $109 million to keep LORAN operating through 2002. So, the LORAN owners won (at least for now), with the LORAN enroute navigation system to remain functional and fully funded until it can be politically and technically phased out safely.

LORAN versus GPS

GPS is definitely here to stay and the entire complement of satellites needed to make GPS reliable and accessible 24 hours a day are now in place. There are, however, FAA restrictions on how GPS interfaces with the airplane that severely limit its use as a navigational tool, especially if you want to hook it up to drive an autopilot. GPS is very accurate, but until the government decides to officially make the signals available to civilians, sudden shutdowns of the system can take place without warning. This has happened, where some airplane owners visited their avionics shop for a GPS problem only to find that the system had been shut off temporarily. At the time of this writing, the governmental gears were put into motion to remove the restrictions and provide full GPS accuracy.

How LORAN Works

LORAN works by measuring differences in the time that two different signals are received from two different LORAN stations. Because LORAN signals are low frequency, the LORAN system, like ADFs, is subject to errors that plague all low-frequency systems. These include day–night, shoreline, and weather effects. These sources of possible errors make it difficult to predict the reliable accuracy of LORAN signals, so for LORAN to be used for IFR approaches, the FAA had to install signal monitors all over the United States to make sure signals are accurate while aircraft are flying approaches. This is the major reason why LORAN approaches have taken so long to be implemented. There are still very few actual approaches in use. More sophisticated LORAN receivers can obtain data from multiple transmitter chains, hence the term multichain receiver.

Most likely, LORAN will be around for several more years, regardless of the phase-out date, so waiting for GPS might not be the best policy if you already have a LORAN stuck in the panel. With the LORAN transmitters operational and the added cost to purchase a GPS, it might make more sense to hold on to the LORAN until the GPS has matured.

Other considerations, which apply obviously to any radio, are warranty and post-sale support. Also, with navigation systems, how much will database updates cost you? Make sure to evaluate the visibility of the unit's display before you make your purchase decision.

LCD displays, for example, can be very difficult to read in direct sunlight. You don't want to find out you can't live with the display after paying out a couple of thousand dollars.

Antenna Installation

Antennas are the sensors of avionics systems, a possible weak link if not given the attention they deserve. Their location and mounting method should be such that no matter how the aircraft is maneuvered, there will be no alteration in the received signal and resultant reduction in quality of the equipment's displayed information.

Four types of antennas are available:

1. For helicopters and low-speed aircraft, use a low-profile bent-blade or bent whip antenna.
2. For top mounting in light aircraft, use a heavy stick-type or blade antenna.
3. For pressurized, high-speed aircraft, use a blade-type antenna.
4. For modern LORAN installations, long-wire antennas require the use of a preamplifier and are no longer used.

The antenna can be mounted on the top or bottom of the fuselage. Make sure a doubler or backing plate is used to maintain the structural integrity of the skin where the antenna is installed and that all mating surfaces are corrosion-proofed and properly bonded.

LORAN antennas must be located at least 24 inches from VHF and HF antennas and 4 feet from DME or transponder antennas to prevent interference. If your installation requires an antenna preamplifier, the distance between the preamp and other antennas should be a minimum of 3 feet. If you can't install the antenna per the LORAN manufacturer's recommendations and you're forced to locate it near noise generators or other obstructions, then skin mapping might be necessary. This involves testing the LORAN while hooked up to an antenna that you can move around to find the best location on the airframe. Once you find the best location, the antenna can be installed permanently.

Some other constraints that dictate LORAN antenna location are as follows:

1. Avoid areas of electrical noise such as strobes, rotating beacons, electrical actuators.
2. Antenna needs physically unobstructed reception.
3. Antenna needs good bonding to airframe.
4. Antenna needs a good ground plane (18 to 24 square inches of unobstructed metal around the antenna).
5. Avoid locating antenna near windshields and nonconductive panels that could build up static charges.

The best location for the LORAN antenna is high on the upper fuselage, but it is also acceptable is on the lower-aft fuselage on the swept-up portion of the tailcone. Although noise levels are normally lower on the bottom of the tailcone, obstructions can be more prevalent, causing possible blanking of the received signal. One thing that might help for a tailcone antenna is bonding the forward nose gear doors on two or three hinge points. Static charges that build up on unbonded doors can act as a barrier, effectively preventing signals from arriving at the antenna.

If mounting the antenna on a contoured surface, you might have to create a flat, conductive surface on which to mount the antenna or install spacers to allow the base of the antenna to bond thoroughly to the airframe.

Because LORANs are particularly sensitive to P-static, treat the antenna with kid gloves. To minimize precipitation static effects, mount the antenna according to the manufacturer's recommendations. Most LORAN antennas are already prepared with anti-static coatings, but if in doubt, ask the installers to apply an additional coat.

Make sure the entire airframe is properly bonded, not just the LORAN installation itself. An ohmic survey of the airframe's bonding should show bonding resistance less than 3 milliohms.

LORAN Installation Tips

Originally a marine navigation aid, LORAN was designed to work best using the consistent conductivity of the salty ocean, but now that LORAN is used over land also, it has inherent problems not associated with marine use. With a low-frequency signal, LORAN is affected by the same atmospheric and electric storm phenomena like thunderstorms, precipitation static, sleet, and rain that cause ADF problems. We can't ignore these problems when installing a LORAN system.

Precipitation static is a cancer that attacks the incoming signal, shakes it up, and shuffles it around until the receiver hears only noise and shuts down. You can avoid precipitation static problems with a proper installation. Here are some ideas to help prevent this annoying and sometimes difficult to eliminate problem:

1. If not already installed, add static discharge wicks to trailing edges of wings and control surfaces to discharge electrical

Static discharge wicks by Dayton Granger.

charges harmlessly into the atmosphere. Replace static wicks periodically to ensure they can do their job efficiently. I don't recommend the carbon-wick type because of their high wear rate. Use the high-resistance rod and metal-pin-style wicks. Make sure the wick base is bonded to the trailing edge and just screwed on over the paint.

2. Use good RF bonding procedures for all equipment. This includes bonding straps, removal of corrosion between components, and application of corrosion proofing solutions like Alodine 1200™ to bare-metal mating surfaces.
3. Use antennas that have precipitation-static-resistant coatings, or spray antennas with antistatic coatings.
4. Torque all antenna fasteners to equal values appropriate to fastener size.
5. If using a wire-type antenna, the wire's insulation must be intact from one end of the wire to the other.
6. If the skin or antenna attachment point isn't metal, install a ground plane.
7. Keep antennas as far as possible from all nonconductive surfaces such as windshields and composite skins that aren't treated with conductive coatings. Composite skins can be treated with a conductive coating that will conduct static charges onto metal surfaces and out to static wicks.
8. Mount antennas at least 24 to 36 inches (36 inches is optimum) from obstructions such as other antennas or airframe surfaces like stabilizers, landing-gear legs, etc.
9. Avoid routing wiring in parallel or close proximity to wiring for pulse equipment like DME and transponders. To reduce potential induction of unwanted noise, cross these wires at 90-degree angles to each other.

The LORAN should be connected into your existing cooling-air setup for your installation. If you don't already have a cooling fan installed and a plenum (chamber that directs cooling air) mounted to the radio rack, see if you can't have it done as part of the LORAN installation. All modern radios must have a cooling fan for reliable operation at high cockpit temperatures. (See Chapter 4.)

LORAN Manufacturers

The following LORAN manufacturers are listed in the Sources section at the end of this book.

Amav Systems

Bendix/King

B.F. Goodrich Flight Systems

H Morrow

Narco Avionics

LORAN CDI Manufacturers

See Sources for addresses of these LORAN CDI manufacturers.

BVR Aero Precision

Mid-Continent Instrument

CHAPTER 14

Global Positioning System (GPS)

Global positioning system (GPS) is the most exciting and accurate navigation system yet devised. The U.S. military created the global positioning system in 1973, but because the FAA wasn't initially involved in the creation of this sophisticated navigation system, the transition to permitting civilian use of GPS has been slow. The FAA, in partnership with Raytheon, is developing the wide-area augmentation system (WAAS) GPS with a price tag of over one billion dollars; however, other agencies have developed a differential GPS (DGPS) system that is considerably less expensive. In 1999, the U.S. Coast Guard's nationwide DGPS implementation team completed the initial stages of installations and are now preparing for the upcoming expansion. So far, seven sites are fully converted from their GWEN (U.S. Air Force Ground Wave Emergency Network) configurations to transmit DGPS corrections. These sites are located near Chico, California; Clark, South Dakota; Driver, Virginia; Penobscot, Maine; Savannah, Georgia; Macon, Georgia; and Whitney, Nebraska. The site at Appleton, Washington, continues operation in its test configuration and the site near Hartsville, Tennessee, operates with a temporary antenna. Currently, the project has plans to construct 59 new sites by the end of 2003. (Table 14-1 shows the sites completed in 2000.)

Table 14-1. DGPS Sites Completed in 2000 (Reference only, check with Coast Guard for latest)

SITE ID	STATE	SITE NAME	LATITUDE	LONGITUDE	RANGE (KM)
876	AZ	Flagstaff	35° 13′ 12″	111° 49′ 12″	450
875	CA	Essex	34° 45′ 00″	115° 13′ 48″	450
847	MD	Annapolis	39° 00′ 36.6″	76° 36′ 33″	*
841	MN	Brainerd	46° 36′ 36.05″	94° 03′ 28.05″	300
849	MT	Polson	47° 39′ 36″	114° 06′ 36″	300
874	MT	Billings	45° 58′ 12″	108° 00′ 00″	400
824	NC	Greensboro	36° 03′ 57.24″	79° 44′ 00.42″	*
851	ND	Medora	46° 54′ 36″	103° 16′ 12″	350
845	NM	Kirtland	34° 57′ 36″	106° 29′ 24″	400
844	NY	Hudson Falls	43° 16′ 12″	73° 32′ 24″	250
823	TX	Summerfield	34° 49′ 12″	102° 30′ 36″	500
873	UT	Myton	40° 06′ 6.66″	110° 03′ 01.32″	400
848	WA	Spokane	47° 31′ 12″	117° 25′ 12″	300

*Sites 824 and 847 not yet active.

Luckily, thus far no one has tried to charge a fee to pilots to use GPS, although it is tremendously expensive to operate. Not only is there tremendous cost in building the system's satellites, but satellite launch and ongoing maintenance is incredibly expensive. Some estimate GPS costs more than $1 billion per year versus the $60 million per year spent on ground-based navigation systems. The planned 7.5-year life span of each GPS satellite is conservative and the satellites should nearly double that life span with an operational capability of ten to thirteen years. You can rely on this: The free ride won't last. Someone will pay either directly or indirectly.

GPS became part of the World Radio Navigation plans in late 1983 after the United States offered the system for civilian use. The first satellite was launched in 1978, and at the time, no one could predict whether the military would permit civilian use of GPS. The future of

GPS looks bright, but it could depend on consumer demand and potential political roadblocks.

Before GPS, the most accurate system we had was DME. While both systems are time-based and can display accurate distance and time-to-station information, GPS can do much more. It is far more accurate and can provide this accuracy in terms of position anywhere on the surface of the earth and also vertical position. In other words, GPS can report your altitude over the earth with pinpoint precision, making GPS a truly three-dimensional navigation system.

GPS is composed of twenty-four satellites in six orbital planes, with four satellites in each plane. All of the twenty-four satellites are circling at 7,500 knots at 10,898 nautical miles high. They complete an orbit every twelve hours while continuously transmitting their position. They are line-of-sight but each satellite can see 40 percent of the earth. When a GPS receiver locks on four satellites, a multidimensional fix is possible. GPS satellites get new data every hour. From any point on earth at any time, vehicles with GPS receivers will have from six to ten satellites in "view." Standard positioning service signals are broadcast on a single frequency in the L-band of the radio frequency spectrum. To ensure accuracy, atomic clocks accurate to within one second every seventy thousand years are part of the satellites' timing circuits.

The advent of GPS could virtually eliminate the need for complex ground-based navigation systems and even relegate FAA's current radar setup to backup status. GPS could provide primary position and altitude data to FAA center controllers quicker and more accurately than the current radar system, but any such transition is still in the future. Even with growing use of GPS, the current VOR/DME system is going to be around for a long time. The FAA would like to phase out the present ADF, VOR, and ILS by the year 2015, but not to start before 2007.

An indication of how popular GPS receivers are becoming is that during 1992, GPS receivers outsold LORAN receivers 13,110 to 5,600 according to the Aircraft Electronics Association. In 1991, the numbers were reversed with 3,300 GPS receivers sold and 8,000 LORAN.

The GPS number includes hand-held GPS receivers and combined LORAN/GPS receivers. These numbers are even greater through the first part of 2001.

How GPS Works

Although the technology behind GPS is highly complex, the basic theory isn't. The satellites are reference points, and position is calculated from distance measurements to the satellites. Four satellites are required for accurate position determination, but as few as three will work in most cases, with high velocity and certain locations being the exception. Receivers that can receive and process eight or more satellites at one time are the most accurate, because the receiver can determine which geometry (which signals of those received from the eight satellites) is the most accurate. Even if all satellites aren't visible to the receiver, altitude sensors tied into the GPS receiver could replace some of the satellite data, reducing dependency on at least one satellite.

With distance measurement accomplished via signal timing, the clocks on the satellites and in the receiver must be highly accurate. Quartz-clock technology made it possible to manufacture a receiver with reasonably accurate measuring capability, up to 0.000000001-second accuracy. Although accurate by most standards, the receiver clock isn't precise enough, but internal computed trigonometry calculations cancel any inaccuracies that occur.

The basic premise behind GPS is the time measurement it takes for a satellite signal to arrive at the receiver. To do this, using received signals from no less than three satellites for an accurate position to be displayed, the satellite generates a pseudo-random code and simultaneously the air-ground (or water-based receiver) generates its own pseudo-random code. When the receiver finds a match for the satellite's random code, then the receiver calculates the time difference between the two codes, which, along with calculations from time differences generated by at least two more satellites, displays location and other calculated information such as speed, time to destination, and so on.

Knowing a little about signal processing will help you decide which GPS to purchase. GPS receivers are advertised as one-, two-, four-, or more channel systems. Many low-budget systems process only one satellite at a time. How fast the data is processed is also a factor of the receiver's capability. During the time processing is taking place, navigation is not occurring, and this can be up to 12.5 minutes for the slowest of receivers. For aircraft, this would not be acceptable, but for a slow-moving boat, maybe. These capabilities continue to change, as manufacturing technology becomes more sophisticated. GPS has become so commonplace that they are purchasable as computer hardware. Prices are low enough that the average hiker can take one along on a trek in the mountains.

Engineers are designing faster-sequencing receivers that track as many satellites as is needed. The drawback is the signal-to-noise ratios (SNR). On a single-channel receiver, SNR is reduced by a factor of four during the time sequencing (switching between satellites) takes place. Designing a single-channel receiver is much cheaper, and it's easier to keep SNR down to reasonable levels. With a two-channel receiver, SNR is reduced by a factor of two, and the more channels, the higher the potential SNR, thus more work goes into trying to "design out" the harmful, high SNR.

An analogy of SNR might be like comparing water flow from the faucet with the amount of contamination it contains. The greater the ratio of water, the healthier it is. The greater the ratio of contamination, the less healthy the water. With avionics, the greater the amount of noise received, the less the amount of quality signal that can be received and processed reliably. This is a simplified explanation, but it should help begin to explain why receivers are so expensive to design and build.

A few detractors suggest that the four-channel units have a downside. While the aircraft is moving rapidly through the air, the unit cycling through the four channels hampers measurement of velocity. For the best of both worlds, two-channel units are recommended. These units process the data from the satellites in "leapfrog" fashion. While one channel is processing the information from one satellite, the next channel is busy acquiring its lock onto the next satellite. With this little electronic trick, no condition-fail message will occur, allowing continuous navigation.

If considering a GPS, don't be dissuaded by the critics. The FAA okayed GPS for VFR navigation, and it still can be used as a backup for IFR and as reference data for certified sensors like LORAN and VOR/DME in some of the newer flight management systems. FAA's TSO C129 is the most recent standard for stand-alone GPS receivers. This will permit GPS to be used for IFR navigation, now that the full complement of satellites are up.

The civilian signals were purposely degraded by the U.S. military to an accuracy of slightly over 325 feet horizontally and 500 feet vertically, but the U.S. government has recently decided to upgrade or remove the degraded tolerances to allow greater accuracy. For point-to-point navigation, however, GPS is amazingly accurate. An example of the types of flights that GPS's accuracy makes possible is the global circumnavigation made by Tom and Fran Towle as far back as June 1990, as reported by *Aviation Consumer* magazine. The Towles flew from Miami to California, to Hawaii, then to isolated, hard-to-find Christmas Island in the middle of the Pacific Ocean. The GPS came in very handy on the 2,446-mile stretch from Cape Verde to Barbados, West Indies. Sixty-one days after they left, they returned to Miami, having logged 150 hours in their Cessna 310, capably assisted by twin panel-mounted Trimble GPS receivers (Trimble is no longer in the aviation business).

GPS IFR Approaches

The FAA conducted testing of GPS IFR approaches using the Aircraft Owners and Pilots Association (AOPA) Beech A36 beginning in late 1992. Assuming that true GPS IFR approaches become reality at some point (perhaps even shortly after this book is published), initial approach certification will be at the more than five thousand airports with existing nonprecision approaches. Until new TSO standards are developed for GPS receivers to be used for IFR approaches, GPS nonprecision approaches will have to be flown using VOR/DME receivers or flight management systems as a backup.

GPS Receivers

Which GPS unit should you select? The list of available receivers is growing, with prices dropping and features increasing. Manufacturers will continue to proliferate as more and more confidence in GPS is realized, by both the FAA and the flying public. At this time, Garmin is the most popular GPS receiver. Many of the preexisting GPS manufacturers have either exited or will exit the general aviation market and concentrate on the marine market.

Installation Tips

The best place for GPS antenna installation is on top of the aircraft's fuselage, as far forward as possible, but no closer than three feet from the windshield. GPS antennas aren't as sensitive to P-static as LORAN antennas. Because GPS relies on line-of-sight reception, any shading from vertical stabilizers or other obstructions can severely limit reception in certain directions. If you're installing GPS in a helicopter, make sure the installer guarantees, in writing, that the rotor blades or the vertical rotor shaft won't affect the system.

Garmin 430 GPS receiver.

GPS antenna by R. A. Miller.

GPS antennas are usually flat, rectangular-shaped devices, but when the aircraft is sitting on the ground, standing water on flat antennas can decrease signal strength. You might consider an antenna with a raised, curved upper surface like Bendix/King's KA91 GPS antenna, or the antenna could be mounted at a slight angle to allow moisture to run off. Although any antenna should be bonded properly to its mounting place, it's not as critical with GPS antennas, and good reception has been demonstrated without the extra effort of antenna bonding. Please don't take this as license to not bond the antenna. Bonding also protects the antenna from static and lightning damage.

Other than the above, installation procedures are similar to any radio. If rack mounted, make sure the installer mounts the rack securely to the side rails using all four mounting fasteners. The harness exiting from the rear of the radio should be supported to prevent chafing and stress on connectors. The GPS wiring harness and coaxial cable should be routed away from heavy current sources and raw RF lines coming from other radios like communications, DME, and transponder.

GPS Manufacturers

These GPS manufacturers' addresses can be found in the Sources section at the end of this book.

Arnav Systems

Bendix/King

Garmin

II Morrow

Mid-Continent Instrument (GPS CDI)

Hand-held GPS Manufacturers

See Sources section of this book for the addresses of these hand-held GPS manufacturers.

Evolution

Garmin

Sony

CHAPTER 15

Autopilot Systems

Long flights and an autopilot can be good companions. Fatigue can easily contribute to the pilot drowsing off during a long flight. Being able to engage the autopilot is like having a copilot and provides some relief from hands-on flying. This is especially true when flying single-pilot IFR or having other responsibilities such as planning the next waypoint or plain physical discomfort from hanging onto the wheel for so long. The danger lies in the pilot relying on the autopilot to be there at all times. What happens when the pilot was up until 0300 hours and had too much to drink? Let's take a closer look.

Two hours into the flight, the pilot's head starts to droop. Too tired to shove a magazine into a door pocket, he throws it onto the glareshield. With the autopilot engaged and the plane flying straight and level into the dreary night, it would seem that all would be right with the world; however, this isn't the case. Unknown to those slumped comfortably into the soft upholstery of the passenger seats, the pilot's magazine has slowly vibrated off the glareshield and has accidentally disconnected the autopilot. The autopilot warning automatically initiates but is not heard or seen. With the autopilot no longer on line, the plane begins a slow but steady downward spin to

Century 41 autopilot.

the ocean floor. The pilot, having drifted into a deep sleep, finally hears the warning and immediately realizes something is drastically wrong but is disoriented and confused.

Only seconds remain between the plane and the water. Finally, adrenaline kicks in; the pilot grabs the control wheel and pulls it straight back, but with the plane already spiraling downward and now upside down, the plane is actually forced to plummet even more out of control. The wings experience G-forces far beyond their engineering design. Fracture points rapidly rip across the metal skin. The wing spars begin to surrender, then give way, snapping like twigs. Formers and stringers crumple like aluminum foil. Moments later, the plane strikes the water's rock-hard surface.

It would be several days and nights before the crew and passengers are recovered from their watery graves.

Yes, fate and stupidity conspired against those that took that deadly flight. First, the pilot did not have enough hours of soloing to safely make the trip. The weather was bad; it was dark, and a storm was looming over the ocean. Second, a series of totally unexpected and seemingly insignificant factors led to this accident. A pilot already tired and exhausted from being up too long the night before should not have been piloting a plane. Either too lazy or too tired, he throws a magazine on the glareshield. With only five thousand feet above sea level and spiraling downward, it was too late for those on board. Let us be clear: This scenario is not the failure of the autopilot. This tragedy was caused by the ignorance and complacency of the pilot, but it just as easily could have been a maintenance oversight. To date, no machine can be made fail-proof to second-guess human error.

What happened here? First indicators are that it was pilot error, but supposing it was not. Closer examination might find that there was mechanical failure caused by a lack of good maintenance procedures. Autopilots are not boiler-plated and are just as susceptible to human error as any other piece of equipment. Everything from the interfacing harnesses, primary and control cables, servos, gyros, and controls have weaknesses that will fail if not inspected regularly. There is a strong tendency to blame the pilot, and many times that might be true, but don't be too quick to assume that one of us failed to do our jobs. We must have confidence in ourselves, but not to the point of ignoring procedures and commonsense possibilities that could add to potential failures.

Aircraft autopilots were designed to provide both a physical and mental break during long flights. However, even with an autopilot, unless it is a two-man crew, the pilot should not leave the cockpit or take a nap, especially when the passengers' safety depends on him or her. Anything can happen and as Murphy's Law states, it usually does. Even the best autopilots are subject to unforeseen failures. The scenario depicted above would be rare, but not impossible. If these situations didn't exist there wouldn't be the need for a pilot. It could all be done with electronics and a few muscle servos. In 1998, in San Diego County, a test was conducted with success that involved cars equipped with sensors and computers that would allow safe automatic piloting of the family car along the highway at freeway speeds. Theoretically, the driver would not be needed until exiting the express lane, but should the autopilot fail for some obscure reason it might be a good idea to have a human driver on board to regain control. Supposedly, a fail-safe system would maintain the car in a straight line until the driver could grab the wheel.

As any good pilot knows, before any flight there should be a walk-around to inspect for any discrepancies in any safety-of-flight items. This would include rivet patterns, oil leakage, antenna damage, tire inflation, proper surface movement, blade damage and obstructions in the Pitot tube. Another point of inspection is to confirm that the static test tape was removed from the static ports. (This should be a large X using high-profile colors such as black or red.) When lives depend on the status of the plane, it is never too much trouble to open the access panels and take a quick but careful look at cable tensions and inspect for broken pulleys. The pilot is the last line of defense between a smooth, clean flight and disaster. No matter who screwed up, the pilot has the final responsibility in preventing a tragedy. To be fair, visual inspections may not determine if cable tension is per specification, but will determine if the cable is extremely loose or if the cables or pulleys show signs of excessive wear.

With the pilot is sitting in the cockpit, he or she should check for free control wheel movement, seat security, throttle restrictions, instrument and radio panel structure, plus a detailed operation of the radios to confirm transmission and reception. There will be a lot more to check, including running the engines and checking oil pressure, temperatures, and compasses (whiskey and gyro). The list may seem long, but there are no shortcuts; any deviation from a step-by-step checklist can lead to disaster.

Once each step of the inspection is complete, except for ground testing the autopilot, the plane should be ready for flight. This plane in the following scenario is equipped with a Century 2000, a very intuitive and straightforward system. The following virtual flight is a narrative and will not cover all the features of the Century 2000.

Warning! After the ground test, do not leave the system turned on; it must be turned off prior to flight!

Once airborne, the autopilot can be engaged by depressing the ON button. The system will default to the heading mode; therefore, the heading should be set to the desired course on the directional gyro (DG) or the horizontal situation indicator (HSI). If different from the course to be flown, the plane will immediately start the turn. Engagement will be annunciated with an illuminated autopilot indicator light. Heading (HDG) and attitude annunciator lights will also illuminate indicating normal operation. The aircraft's attitude during climb-out will be maintained, but if a higher or lower speed is desired, the modifier button should be pushed to set for a higher or lower angle of climb. For all practical purposes, for every second the up/down modifier button is held, the angle will change at a rate 7/10ths degree of "deck angle."

For example, if climbing to an altitude of 4,500 feet, without an altitude preselector, the ALT button will have to be pushed when the altimeter indicates a level of 4,500 feet. The aircraft will immediately start to level off. If there is any overshoot, the plane will return to the altitude at the time the ALT button was pushed or 4,500 feet. When the aircraft's altitude is at "zero error," the plane will hold that altitude. *Note:* for detailed operations, please review the manufacturer's features or pilot's manual.

Just think, in the wink of an eye, with the simple touch of a button the pilot has turned the piloting over to the autopilot. The pitch synchronization to the aircraft's attitude provides a very smooth, almost imperceptible transition. Turning the autopilot off is just as easy. Depress the OFF button and a warning will sound and the autopilot annunciator will flash for five seconds and then go out. All modes will automatically revert to HDG and ATT in readiness for re-engagement.

To set the VOR heading, dial in the correct VOR until the OBS needle centers. The plan is to follow the NAV signal. An intercept angle of up to 45 degrees can be performed using the standard DG and up to 179 degrees intercept using the HSI. The NAV button should now be depressed with the plane engaging and navigating the VOR signal.

After navigating several VORs, it is now time to start a preliminary let-down. It is far too early to turn off the autopilot, just yet. Push the down modifier button and hold it in for several seconds to set up a descent rate of approximately 500 feet per minute. When this is done, the autopilot automatically reverts to the ATT (attitude) mode taking it out of the altitude hold mode.

The plane is now descending to intercept the ILS and switch back to the HDG (heading) mode. After a few heading changes, the pilot switches on the proper instrument landing system (ILS) frequency on the NAV receiver and prepares to push the APR (approach) button when the localizer needle is less than 90 percent of center. Of course, it was noted that there was a valid GS and NAV flags and that the GS light came on showing that the glideslope was armed.

As the plane comes across the outer marker, the checklist is conducted for landing. For the next few moments, sit back and monitor the autopilot approach. Not all the configurations of this autopilot have yet been discussed, but that isn't the purpose of this exercise.

The heading mode is quite possibly the most-used mode on the autopilot. Simply adjust the heading bug to any selected new heading and the plane automatically swings to that new setting and flies smoothly with a sense of solid control. As the plane assumes the new heading, the plane will bank to the preprogrammed bank and roll rate limits.

Autopilots have a wide availability of potential, from basic wing levelers to total electronics that will do everything except park the plane. Essentially, the main purpose of an autopilot is to relieve the pilot of hand flying by sensing immediate changes in the plane's axis and correcting to level flight using servo mechanisms. Also, if the pilot wishes to make a change, he or she can do so by simply pushing a button, without grabbing the control wheel. All of this is simply an electromechanical extension of the pilot's hands and arms.

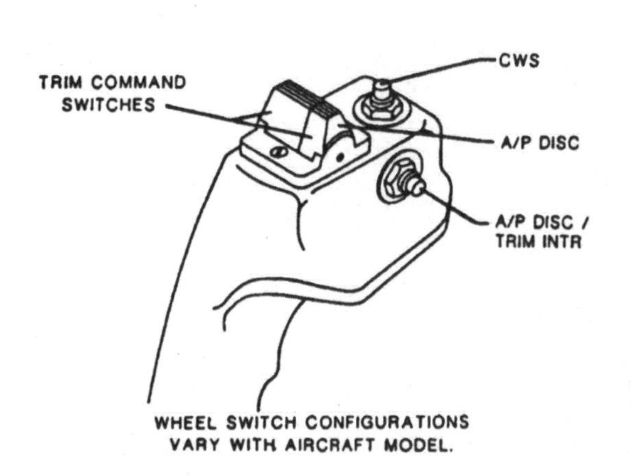

Pilot's control wheel—typical autopilot switch layout.

The technician and pilot must have greater insight of weak links within any given avionics system and discover how to prevent unwanted and unwarranted failures. Allowing for human error, equipment failure, and design inadequacies, obvious weak links are the control cabling in both the primary and secondary (bridles) system. However, before blaming autopilot problems on too-loose or too-tight cables, bench-test the pulley brackets, tensioners, and gyros to ensure they are within operating specifications.

There are two basic design approaches to autopilots; the Century Autopilot described earlier is a rate- and position-based system, whereas the Allied Signal or King is just position-based. The price difference between the two is considerable. The Century system can be purchased for as low as $5,000, using a roll servo, whereas the King starts somewhere about $14,000. Not to get into a price debate, the point is that one system is set for a different customer market than the other. One nice thing about the Century is the ability of the customer to create a system one upgrade at a time, in modular form. The dominant autopilot design is "attitude-referenced positioned." This type of system continues to lay the standard for flight control systems. The downside to a pure position autopilot is the inability to compensate for bank overshoot, slow reaction time to commands, and the wide variation in control reaction from one aircraft to another. Recent solutions that mirror the commercial aviation industry

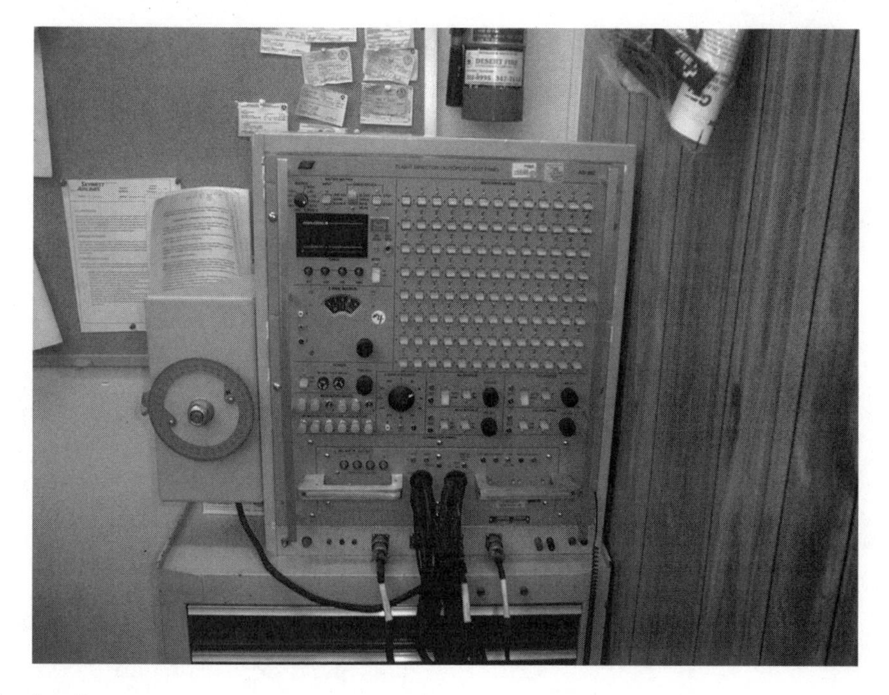

Autopilot test set.

are a compromise of both the rate and position sensing. This is made possible only by the use of complex electronics.

Regardless of the manufacturer, all autopilot systems require some form of control feedback and followup to limit control surface velocity or deflection for any given error. The problems come in when conditions vary that affect positive and negative pressure on surfaces. This is where cable tensions can create some weird situations. These variations occur when the plane is at cruise speed or the opposite, slow speeds. Weather conditions are also factors that must be considered. The autopilot must be able to compensate accordingly. The computer, gyros, and remote sensors provide the feedback loops that keep the ship flying level.

Autopilot Gyro

The gyro is one of the critical components of the autopilot. Sensing relatively minute changes in the plane's attitude, the gyro's internal sensors provide dynamic information to the autopilot computer for immediate correction. This data is received, processed, and acted upon by sending signals to remote-control surface servos, returning the aircraft to its prescribed course.

Autopilot flight director in center of pilot's panel.

Regardless of the plane's location while in flight, the gyro will attempt to stay erect in relation to the earth's horizon, sending compensating electrical signals only when the plane's attitude alters. Of course, there are obvious internal and mechanical restrictions. For example, should the plane bank sharply, the gyro will roll against the internal stops, staying there until the plane returns to normal flight configuration. The electrical signals, in order to do any good, must have muscle. The servos (reversible-variable speed) serve that purpose by directing through the flight computer the electrical power to drive a remote servo, which then adjusts the control surfaces to correct deviation.

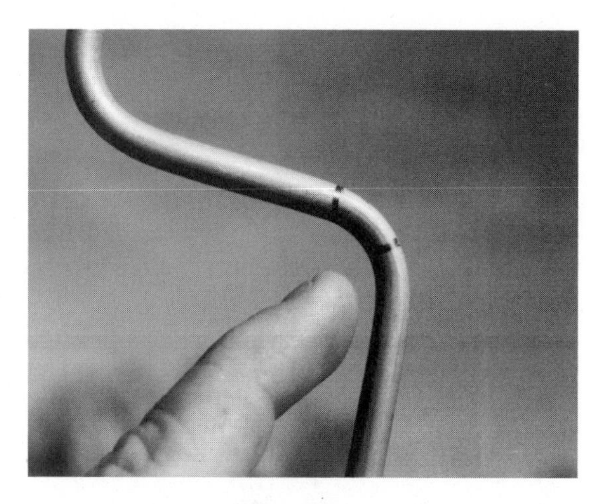

Check for accurate tube forming on solid plumbing used behind panel.

The main gyro used in the autopilot system is the flight director. It can be considered to be an overseer, surveying all sensory input for small changes, then reacting, sending out signals to the computer to correct to programmed command values. Additionally, specific alerts or warnings may be displayed visually or audibly to warn the pilot of impending situation conditions.

Being panel mounted, the gyro requires a secure, trouble-free installation to prevent future mechanical and electrical failure. There are several possible installation methods, but all require that the installer allow for cosmetics, forward G-force of the plane during landings, vibration during engine operation, constant handling and clearance for harnesses, and plumbing. Vendor installation manuals will try to illustrate their recommended methods for installation; however, this may not satisfy a given custom installation to be performed on a future modification.

Relatively small amounts of electrical current flow through the nervous system or control circuits of an autopilot. It potentially can be affected by outside influences such as poor installation with sloppily routed power wiring, antenna problems, and mechanical failure. It follows, then, that the smallest error, usually human or electronic, can affect the accuracy and operation reliability.

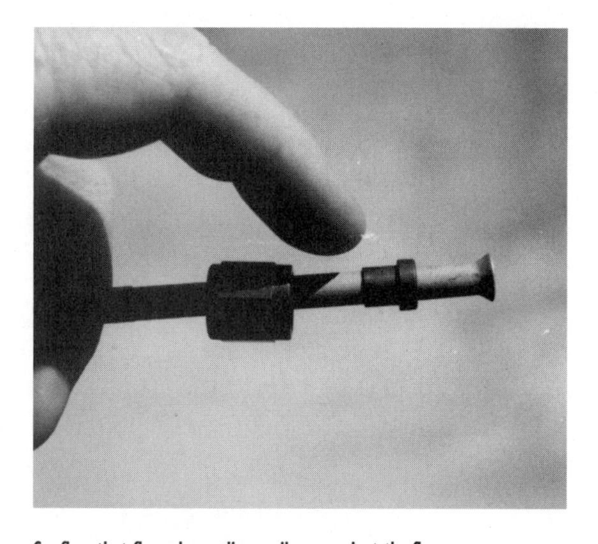

Confirm that flare sleeve slips easily up against the flare.

A correctly formed flare will not leak; leaks can cause some strange autopilot problems.

Engineering design attempts to prevent any error by the autopilot, but if the gyro is ignored this is all for nothing. This error can occur should the gyro fail from bearing race breakdown, or airflow restrictions caused by kinks or contamination. Failures may also occur from incorrect gyro mounting angle, over-clamping of hoses, high G-forces (a drop of only a 1/4-inch can result in up to 20 G-forces), poor instrument panel shock mounting, excessive heat, or electronic malfunction.

The gyro has a history of operating flawlessly for thousands of hours, without any significant problems, if not banged around. This is why the technician must shoulder much of the responsibility to provide the necessary inspections and maintenance to insure continuous, trouble-free operation. As noted earlier, gyros do fail, but shouldn't—at least, not during a flight, and not if properly maintained. As part of scheduled inspections, the potential failure points should be checked and logged for the pilot/owner to make maintenance decisions.

Historically, the gyro won't go in for repair or evaluation until some form of failure occurs. Previous documentation from FBOs around the continental United States are showing premature failure from internal wear occurring after one thousand hours, or more. In these cases, bearing-to-race friction wear, inner or outer gimbals' surface erosion, and heat play a major role in internal component failure. Because proper operation depends on an adequate flow of air pressure, minute particles of dust, tobacco smoke, and other contaminants can clog the filters and reduce efficiency. Regularly scheduled inspections (every one hundred hours) are highly recommended to evaluate filter contamination. The environmental conditions are harsher in some locales than others, and filter replacement may be necessary. Because inspections of the primary cable are required at this time, it is cost-effective. For some reason, inspection of the control cables is overlooked during these inspections. Filters should be replaced, not cleaned. *Caution should be exercised during this process; all plumbing lines should be capped during instrument removal.* Realistically, this isn't always the case. Technicians and mechanics alike will quickly remove the indicator and run it to the stockroom for shipping. Too many times the plane will sit with the lines open until the new part arrives.

The bearings (jeweled movements) are manufactured from artificial rubies that may shatter should they be subjected to sudden shocks. Virtually all indicators should be handled, to use a well-worn quote, "like eggs." Actually, eggs can survival tougher handling than many instruments. Electrically powered indicators have jeweled movements; only some engine gauges having simple brass-to-brass bearing surfaces. The fact is, even if the design of the instrument is second nature to the technician or mechanic; when handling instruments, it would be wise to treat them all like they could burst open and spew out a toxic gas!

Maintenance

Look for inaccurate indications, precessing, slow erection, spinning, or tumbling. Precessing will be seen during power-up operation, but the spinning or tumbling usually occurs during power down. Damage to bearings earlier than the previously mentioned one thousand hours can be attributed to handling prior to and during initial installation or during maintenance.

Items requiring routine inspection and recommended maintenance on a yearly basis or every one thousand operating hours, whichever comes first:

1. Remove capstan assemblies from servos, check clutch settings, and readjust as required to meet vendor specifications.
2. After reinstallation, set bridle cable tension (also recommend checking and readjusting, as required, the primary cable system).

3. Remove the gyros, change filters, blow airlines clean with nitrogen gas, bench-check for precessing and check records to determine if scheduled overhaul is necessary.
4. Perform ground and flight-tests to vendor specifications and test procedures.

Many autopilot computers are remote mounted and have the potential of being subjected to detrimental environmental conditions that may degrade performance, even to the point of causing total failure. Moisture-carrying contamination and other potentially damaging materials could eventually work their way through, penetrating the sensitive innards of the computer. This may result in the moisture-laden contamination shorting across the exposed contacts on the printed circuit boards, and ultimately destroying the ICs or other delicate components.

Enemies of the autopilot are heat, cold, vibration, poor routing (sharp bend radius, insufficient harness support, sharp edges, etc.), and location. The autopilot computer should be located at least 12 inches or more from the strobe power supply. It should be at least 1/4-inch from other boxes and structure to prevent physical interference.

For hard mounting of the computer, use mounting fasteners capable of withstanding the high G-forces of takeoff and landing, especially during those hard landings. If secured to honeycomb shelving, fasteners should incorporate a through-shelf design. When mounting the computer, confirm that the base of the unit lays flat. It should make simultaneous metal-to-metal contact at all four mounting points. Apply electrical and RF bonding procedures to achieve a resistance reading of less than .030 milliohms between component and structure. Remember, final fastening and resealing must occur within twenty to thirty minutes of applying Alodine 1200™ solution.

Allow a service loop for easy removal of the connector. Cables being too short may more serious then being too long; support the harness directly after the connector, within 8 to 10 inches of the computer body (right where the service loop begins). Wires exiting the connector must be clamped to prevent damage to the pins and sockets within the connector.

Besides the computer and the Flight Director (FD), there is the control head (controller) and annunciator. The control head is usually* located in the pedestal, just aft of the throttles and in most cases, requires only two angles to support the rack that will hold the controller. If there is a removal mounting plate on the pedestal, it will have to be removed and cut to match the unit. Supporting angles are required to be riveted to support and secure the mounting rack for the controller.

Be certain the connecting harness is clamped securely and isn't in close proximity to any of the surface control cables, throttle controls, or wiring. The annunciator should be located in easy view of the pilot, usually just above the flight director, and below or in the glareshield. Mounting may require a mounting rack, but in most cases, it will only require four screw-holes to secure the unit.

Feedback and amplification circuits within the control system of the autopilot have very low voltages that are susceptible to extraneous signals such as fluctuating power circuits and transmitters. Signal wiring must be routed to prevent accidental disconnect or system malfunction. Shielding, twisted pairs, bonding, and other preventive measures help ensure that the installation is secure and reliable. Sadly, these aren't the only liabilities that can potentially degrade the efficient operation of the autopilot. The more expensive and complex the system, the greater the potential for error. As autopilots increase in relative price, the number of commandable axis and options can multiply, resulting in more circuitry, thus more areas for failure.

After installing and completing the installation, operational check the system, but be prepared to observe the cables during operation for takeoff (exit from pulley V groove) from pulleys and capstans. If the pulley is not mounted in direct line with the cable, the pulley might break with the cable possibly jumping off, creating excessive slack, which can be a lethal situation. This is extremely important

*Some controllers may also be mounted in the instrument or radio panel.

with primary cabling. None of the cables should rub the sides of feed-through holes, wiring, structure, or brackets; otherwise, they could easily chafe to the point of fraying and eventual failure.

Primary/Bridle Cables

Most malfunctions within the autopilot system can be attributed to the primary and bridle cables becoming slack, or worse, that they were never set properly during previous maintenance. Normal slackness can result through long-term stretching, or when the plane flies into geographical locations where the temperature is drastically different from the area where the plane was originally calibrated. Symptomatic of this condition is porpoising (derived from our finny friend the porpoise), the rising and falling of an aircraft in a shallow vertical plane exceeding roughly a ten-second period. This is maybe the most common problem experienced by autopilots. For the most part, this can be relatively easy to correct, as the usual cause for this complaint is excessive slack in the primary and bridle control surface cables.

Consider the implications of this effect on the entire length of the airframe. It would become clear that a fuselage could actually shrink or lengthen as much as 5/8-inch, while the control cables, protected to some degree, will change very little. This unequal variation in tension will cause operation problems in the autopilot system by changing the length of the associated control cables, thus confusing the autopilot and its sensors.

Many forces act upon a plane while it's in flight, and if cable tension is out of tolerance, the autopilot will overwork itself as it overshoots its prescribed limits while attempting to compensate for the incorrect travel. This effort by the autopilot to correct for out of tolerance cables can effectively reduce the life of the autopilot system. If the cables could be seen during flight, they would each look like a whip being snapped. First, there would be sudden and excessive slack, then cables would become extremely tight, repeating over and over again. Besides the obvious porpoising that could occur, the excess slack could cause the cable to fly over the radius of a pulley causing it to lose considerable tension, possibly snapping downward and breaking the pulley. Even more possible is the cable could snap off the pulley and create excess, unrecoverable slack that potentially could cause loss of primary surface control and even an accident. Wear and tear notwithstanding, the servos, cable guides, pulleys, and even electronics will be affected as a result of the heat buildup and stress generated by the constant and seemingly relentless operation of the autopilot system.

Ground Check/System Alignment

To make sure that those troublesome malfunctions don't crop up unexpectedly, a basic ground check or a system alignment should be performed. This process would entail aligning the system with the gyro, adjusting the command bars, and determining if everything is properly centered. In checking performance, a performance sheet or preflight checklist must be used that is specific to that aircraft. Typically, things to scrutinize on the checklist are the failure monitors associated with their respective systems; these can be computers and heading flags, warning monitors (important), and acceleration monitors.

Essentially, what we're looking for is anything that's related to self-checking the operating system. These factors should be thoroughly checked for any abnormalities. If everything appears to be normal, an autopilot engagement check would be initiated. Once the autopilot has been engaged, determine if the control wheel goes left and right with the respective left and right commands. Determine if the wheel will show a tendency to respond with up and down movements after receiving the proper pitch commands from the autopilot.

Confirm that the autotrim is trimming in the correct direction. The effective method for accomplishing this is to allow the control wheel to relax, provide a pitch-down command, allow the elevator to go downward, and after about 3 to 5 seconds the autotrim should also run in a downward direction. Follow up with same test in the opposite direction by applying the pitch-up command confirming that the elevator does come up, somewhat, and that the autopilot, after 3 to 5 seconds, trims in the upward direction. This completes the check for proper trim phasing.

Next, check the autopilot override. This will confirm proper trim clutch settings—that they're set up where the pilot can override the autopilot, if need be. Force the control wheel left and right, overriding the autopilot; this will effectively stop the autotrim from running. Use the resultant elevator movement as an indicator of the pitch up and down. The same holds true for the yaw damp system, if installed.

A yaw command will have to be simulated for the yaw test (does not include Century autopilot; it can only be tested during a maintenance operation).

Rotate the attitude gyro approximately 25 degrees, which should simulate a roll crossfeed to the yaw damp system. *Note:* The procedures outlined within these paragraphs are, for the most part, generally applicable to several autopilots, but because every system has a different method of supplying yaw commands, reference will have to be made to the applicable vendor's manuals and specifications. After simulating the yaw damp command, determine if the rudder pedals are traveling in the proper direction and if the yaw system (clutch) can be overridden, thus stopping the travel.

Determine if the autopilot system can be disconnected with autopilot disconnect switch. There are two different types of control wheel disconnects:

1. The autopilot disconnect—with this one, the remote controller will be lost along with the pitch and roll.
2. The trim disconnect—this will disconnect autopilot, which will also indirectly disconnect the trim system, such as in the case of trim runaway. (Because different autopilot systems vary in operation, this may not apply, particularly to the Century system.)

Runaway may occur in much older King autopilot systems if a transistor fails because of shorting within the trim servo while power is applied. This is a rare situation and only applies to autopilots manufactured during the 1970s, but the switch is there, if needed. These disconnects will use either a one- or two-detent button that performs two functions (older models used two detents). The first function will disconnect the autopilot and the second function is the trim disconnect which will disconnect the trim; if the switch is a one detent, both disconnects take place simultaneously.

The entire autopilot should also disconnect when the compass system is shut down (loss of compass valid) while in the NAV mode. With some autopilots, a horn and light should normally activate during disengage. On earlier systems, depressing a test switch could disengage the autopilot. Contact the vendor for a modification to correct this condition. In any case, refer to the pilot's operating handbook for correct operating procedures for other systems.

When making a ground alignment, the command bars shall be positioned (read adjusted) to about 20- to 30-degree angle to the center of the indicator (as viewed by the pilot, this will vary according to the pilot's height). The pitch axis can also be adjusted for the same viewing angle (slightly above center). The roll can be final adjusted after the autopilot has been aligned in both pitch and roll and then the command bars can be tweaked to not show any sign of left or right roll.

Autopilot trim is next and must be evaluated in flight by the pilot interfacing the plane with the earth's horizon, the altimeter, the rate of climb, etc., to determine that the plane isn't gaining altitude, that it is, in fact, flying straight and level. Pitch trim the aircraft to determine that it can maintain altitude without the pilot applying any force on the control wheel. Prove out roll alignment by evaluating the distance between the wing tips and the horizon; if the space is unequal, equalize while monitoring the attitude as adjustments are made.

To do this, look at the turn and bank (T&B) to see if there is any additional roll. Additionally, check the wing tips with the horizon. Follow up by adjusting the roll trim. When checking rudder trim, don't reference the turn and bank for rudder trim checks; look at the magnetic compass or slaved compass, if there is one. If neither is available, use the whiskey compass; it acts as a natural level. The compass may be used to determine if the aircraft could be in a slight turn, left or right. Watch for slow movement in the compass; if any, adjust the yaw trim to compensate. This is the final and most essential check in determining if the aircraft is in proper trim while in flight.

Now that the trim is complete, the autopilot engagement can be initiated. Again, look for wings that are level, then engage altitude hold and monitor for any noticeable bumps, porpoising, or erratic movement. At this point, also check to confirm that the chosen altitude is holding; if so, the heading can be now engaged. Keep a close watch for heading tracking of the heading bug. Check for noticeable left- or right-wing down in tracking the heading bug or keep an eye on the heading bug to see if there's a noticeable line, left or right, as opposed to the lubber line on the compass system. If there are any problems in this area, such as a wing lowering itself, this might indicate a slight gyro rotational alignment is needed.

Engage the autopilot (the aircraft should be completely in trim), and check for any pitch problems that may occur on engagement. Also, check for any wing right or left low conditions. There should be no bumps; if there are any, this indicates that the pitch centering may need to be adjusted. If a wing was lowered, this would be good time to adjust the roll null centering, until both wings show level. At this time, altitude hold can be engaged. There shouldn't be any change in altitude on the altimeter after engagement, nor should there be any noticeable bumps. If this looks good, the heading can now be engaged; the plane should track heading within manufacturing specifications.

If the plane has trimmed properly, and it should be by now, and there are remaining problems, there must be an error in the flight control system; specifically, it would be an error in the heading portion of the system. The heading output of the directional gyro may not be within tolerance, or the computed output of the flight computer may not be accurate. These should be the only two possibilities for the autopilot not to fly the heading bug, except for wiring and EMI interference.

When intercepting the NAV radial, a 45-degree intercept would be ideal. To engage NAV mode on the flight control system there should be a NAV armed annunciation and the system should track as the plane comes closer to the selected bearing. The autopilot will then capture, smoothly roll, and track the NAV needle. The flight control system should go into the track mode and follow the right and left needle within manufacturer's specifications. If this doesn't seem to be the case, the ideal thing to check is the NAV receiver.

The tracking of any signal by the autopilot is only as good as the signal received and output. To monitor NAV receivers, first decouple the autopilot and visually watch the NAV needles; monitor for any erratic movement while holding a pre-established heading. Look for any erratic movement; the needles should be steady and shouldn't drift slowly left or right while the heading is held. A small warning: Don't test the middle marker during approach with couple engaged on older systems as the autopilot will respond to the resultant gain reduction.

Installation is only a part of the overall reliability of the autopilot system. Follow-up and good maintenance are key to continued failure-free operation. It all becomes a lot easier when original assembly problems have been eliminated as the potential source of future failures. This is where the technician's real responsibility for the survival of the autopilot begins.

It should be understood that regular inspection of all brackets, pulleys, cables, equipment, and indicators is necessary to ensure that there will be no unwarranted failures. Again, this is especially true of cable tension in the both primary control surface and secondary autopilot system. Blending the cabling system with an understanding of system operation and aircraft dynamics is perhaps the most critical part of doing any preliminary troubleshooting. This is the role of both the plane's crew and ground support personnel of technicians and mechanics. Let's fact it, in the end, everyone is responsible for the safe operation of the aircraft.

Keeping Autopilots Going Straight and Level

Autopilots can be as simple as a turn coordinator that provides information to drive an aileron servo for a simple wing-leveling system or as complex as an integrated flight management system that costs hundreds of thousands of dollars. The types most commonly installed in light airplanes are the wing-leveler and/or more sophisticated systems with altitude hold, glideslope capture, and complete NAV tracking capability.

An autopilot basically consists of the following components:

1. A controller with annunciators and buttons for pilot operation (what you see in the panel).
2. An autopilot computer/amplifier, which manages and controls the signals that travel from the attitude-sensing devices such as attitude gyros, barometric sensors, and turn coordinators to the servos.
3. The servos, which translate the autopilot computer signals into action by moving the control surfaces, ailerons, elevator, rudder, and elevator trim. Servos are attached to control-surface control cables via bridle cables, which are short cables that run from the servo to the control-surface cable.

You've no doubt come across the term "axis" when reading about autopilots. Sometimes a wing-leveler autopilot is called a one-axis autopilot, and an autopilot that has aileron and elevator servos is called a two-axis autopilot, or sometimes three-axis if the autopilot controls elevator trim, too. Does that make an aileron, elevator, elevator trim, and yaw damper autopilot a four-axis autopilot? Perhaps, but the reason I mention this is just to warn you that the axis reference is by no means standardized, and for accuracy, you should find out what an avionics seller is talking about when they refer to autopilot axes. Don't just assume they mean aileron, elevator, and elevator trim servos when they talk about a three-axis autopilot.

Gyro Erection Failure

One of the critical gyros is the attitude indicator; it is the only instrument on the panel that provides a clear and immediate picture of the attitude of the aircraft while in flight. Without it, the pilot must create a virtual mental picture of the flight attitude by combining information from several other instruments, such as the T&B (turn-and-bank) or turn coordinator, the airspeed indicator, altimeter, and the vertical speed indicator). These indicators provide only partial, indirect, and, of course, delayed information about the flight attitude.

Most GA (General Aviation) airplanes are equipped with an air-driven attitude indicator powered by the aircraft's vacuum or pressure system. The heart of the instrument is a gyroscope, not unlike a child's toy gyro; however, instead of a string, it relies on a steady input of filtered air to keep it spinning. Should the air supply falter, the gyro fails and so does the plane, if on autopilot.

The "artificial horizon bar" on the face of the attitude indicator is mechanically coupled to the gyro spinning inside, so it remains parallel to the natural horizon, regardless of the plane's attitude (within normal limits). The instrument's little symbolic airplane is fixed to the instrument case, so as the plane changes attitute, the case moves, but the gyro does not. This provides a fixed reference between the plane and the gyro. Therefore, the relationship of the horizon bar to the symbolic airplane is the same as the relationship between the natural horizon and the real airplane.

As a pilot, you should keep an eye on your attitude indicator's initial erection process; it's your best clue to the gyro's internal health. As gyros get older and more contaminated, the instrument may not erect quickly; the pendulous vane pivots maybe contaminated and sticking. If the oscillations don't damp out shortly thereafter, the rotor may not be coming up to full rated speed (due to worn or contaminated spindle bearings, low vacuum, a pinched air hose, or a dirty air filter). On the other hand, if the horizon bar hardly oscillates at all during gyro spin-up, the gimbal bearings may be worn or contaminated. In any case, do something about it! Get it to your favorite FBO immediately and get it overhauled or exchanged.

Why is this so important? When flying on autopilot, a failure of the attitude gyro or its vacuum power source will cause the autopilot to try to follow the slowly precessing attitude gyro as it spins down. In most cases, the autopilot has no way of detecting the failure, so it continues to follow the failing gyro until the aircraft assumes such a severe, unusual attitude that it exceeds the autopilot's servo torque limits. Many inflight structural failures and loss-of-control accidents are traceable to gyro failure while flying on autopilot. In most cases, by the time the pilot became aware of the problem and tried to recover manually, the aircraft attitude was so severely upset that recovery was not successful prior to structural failure or impact with terra firma.

That's why it is so critical for pilots of autopilot-equipped aircraft to maintain a vigilant instrument cross-check while the aircraft is on autopilot, and also to provide systems redundancy so that the attitude gyro won't fail in the first place.

It is for this reason that single engine aircraft that fly serious IFR should have a backup vacuum system or electrical gyro. The plane should be provided with dual electrical and vacuum systems. It is not worth your life to cut corners.

Autopilot Gyro Installation Tips

Being panel-mounted, the gyro requires a secure, trouble-free installation to prevent future mechanical and electrical failures. The installation must allow for easy viewing by pilots, G-forces during landing, engine vibration, clearance for harnesses and plumbing, and ease of access to prevent shocks to the unit during removal and installation. Vendor installation manuals illustrate recommended installation methods; however, this might not satisfy a particular custom installation.

More expensive flight management systems use remote gyros, which can be larger and thus provide better information to the autopilot. However, it's unlikely you'll see one of these in a light, piston-powered-airplane autopilot installation.

Autopilot Installation Tips

When considering a new autopilot, two of the central companies you'll probably be looking at will be Century Flight Systems and S-TEC. Both offer modular systems that can be expanded upon, starting with a simple, relatively low-cost, single-axis, roll-control system and building up to a sophisticated, altitude-hold, navigation-coupled flight control system. My recommendation is to go with a modular system if you plan to keep the airplane for a while and want to add to its capabilities as you can afford to. Less expensive nonmodular autopilots are available, but they limit you to whatever you purchase at the time and can't be added to later on.

Enemies of the autopilot are heat, cold, vibration, poor wiring routing, and incorrect location. Avoid having the autopilot computer located next to the strobe power supply; the computer should also be at least 1/4 inch away from other black boxes and any nearby structure. There should be enough slack in the wiring harness connected to the computer to allow easy removal, but not too much slack that the wiring just flops around and rubs on nearby metal. The harness should be supported 8 to 10 inches from the connector at the computer to prevent too much slack in the line.

Autopilots to be installed in an aircraft must be STC'd and that means the location is fixed per the STC. If you are having an autopilot installed for the first time in your airplane under a new STC and there isn't a standard place to put it (like the factory location for the Century autopilot on Pipers), consult with your installer about where you want the control head and annunciator lights located. Sometimes the ideal spot for the control head is on the center pedestal, behind the engine controls. Your installer might have to manufacture a bracket for that location, and care must be taken to prevent autopilot wiring from interfering with engine controls. The annunciator panel should be installed where you can see it easily. Many installers put it just above the flight director (or attitude indicator, as the case may be) but below the glareshield. The point here is, your input is important because you're the one who will be flying with this installation, so get involved during the process so that you'll have a better chance of a happy result.

Feedback and amplification circuits in the autopilot's control system are low voltage and susceptible to interference from extraneous signals caused by fluctuating power circuits and other radios, especially transmitters. These circuits are also referred to as sensitive circuits because they are sensitive to outside interference and must be protected to prevent accidental autopilot disconnect or other malfunctions. Normal use of shielded, twisted-pair wiring and proper bonding of all components will help assure the autopilot will be efficient and reliable, but keep in mind that the more complex the autopilot, the more potential areas of sensitive circuitry are installed and thus the more potential for introducing errors.

Before buttoning up the airplane after autopilot installation, go over all the cabling with your installer and watch the action of the autopilot cables as the autopilot is tested. It's very important that cables remain securely on their pulleys and capstans during autopilot operation, and none of the cables should come close to rubbing sides of feedthrough holes, wiring, structure, or brackets.

If, as is usual, the autopilot computer is installed in a remote spot in the airplane, make sure it is well protected from potentially harmful moisture and excessive heat. Don't just let the installer put it in the most convenient place, find a spot that will contribute to the computer's longevity.

Autopilot Problems

One of the most common autopilot problems is porpoising, which occurs when using the autopilot in altitude-hold mode. Porpoising is when the autopilot causes the nose of the airplane to hunt up, then down, roughly every ten seconds, making any type of flying very uncomfortable. In most cases, porpoising can be relatively easy to correct because the usual cause is excessive stack in the flight control cables and autopilot bridle cables. You can prevent this problem by simply ensuring that your aircraft's cable tensions are kept within proper limits at your annual inspection.

Many forces act upon the airplane in flight, and if the control cable tensions are out of tolerance, the autopilot tends to overwork itself as it overshoots and undershoots its prescribed limits while attempting to correct for the incorrect travel. The autopilot's attempts to correct for loose cable tensions can reduce the life of the autopilot. Servos, cable guides, pulleys, and even autopilot electronics will be affected due to heat buildup and stress generated by the constant and relentless operation of the autopilot.

In one case, I discovered the cause of the porpoising to be a loose bridle cable, but it wasn't because of a lack of tension. The autopilot installer had used undersized rivets to hold a pulley bracket, so one end of the bracket moved every time the cable was tensioned and the system acted as though there was a lot of slack in the bridle cable. This customer was lucky—a simple replacement with stronger rivets solved the problem.

During a climb to a preselected altitude, slack cables will cause problems in establishing the correct altitude. After normally overshooting the preselected altitude, the autopilot attempts to recover to the correct altitude, but because of excess cable slack, it compensates improperly and the airplane levels out at an altitude higher than selected. The more slack, the worse the overshoot.

In another scenario, after a porpoising incident, all cable tensions were checked and were found to be within limits. The autopilot computer and servo checked okay, but the airplane still porpoised. At this point, we had to consider cold soaking as the cause of the porpoising. Cold soaking is where low ambient temperatures cause the aluminum airframe to shrink more than the steel control cables. The colder it gets, the looser the cables. The center section of some large twins can shrink up to a hundredth of an inch in cold air. This new slack in the control cables can confuse the autopilot and its sensors, causing porpoising and other problems. The trouble is difficult to track down because once on the ground in warmer temperatures, everything checks out okay. Some manufacturers compensate for this problem by adjusting control cable tensions to the high end of the recommended setting, especially for longer airframes that fly at high altitudes.

One way to prevent the cold-soaking problem is to have your aircraft's cable tensions set after the airplane has been sitting in a cold hangar all night. Even though you're not duplicating the effects of flying in cold air at high altitudes, the settings will be closer to providing the necessary cable tension than if the tensions are set in a warm hangar with the airframe expanded.

A rule of thumb used for preliminary tension checks calls for bridle cable tension at 50 percent of the primary control cable tension. Don't use this as a final setting; instead refer to the specification of the airframe and/or autopilot manufacturer. Once the airplane warms up again, it isn't a bad idea to recheck the tensions to make sure they don't end up out of tolerance as the temperature changes.

Once tensions are set, perform an operational test to evaluate autopilot function. Although pilots rarely experience problems with overtight cables, one symptom can be oscillation effects similar to porpoising, but with a more pronounced but shorter period (usually less than ten seconds) between up-and-down cycles.

Slack cables can also cause problems in the lateral axis like overshooting of bank angle when using the heading bug to change heading while on autopilot. As the autopilot computer senses that the desired bank angle has been achieved (to give a standard-rate turn), the computer tells the roll servo to zero-out the ailerons, which should hold the airplane at the desired bank angle. The slack in the cables, however, causes a greater bank angle than that needed for a standard-rate turn.

The attitude indicator, sensing the error, will transmit a correcting signal to the autopilot servo. But the slack in the cables allows the servo to drive the ailerons to the incorrect position, once again followed by the attitude indicator's repeated sensing of the error. As this is a closed-loop system, the action continues, resulting in what is commonly known as "wing rock and roll."

Other elusive autopilot problems that cause porpoising and sloppy pitch and roll control or even undesirable Dutch rolling are usually traceable to contaminated connectors in the autopilot servos and computer. Assuming all cable tensions are correct, the next troubleshooting step is to carefully check all electrical connectors in the autopilot system.

One of the preventive measures I recommend to prolong autopilot health is periodic maintenance on the servos. Every year or one thousand hours, whichever comes first, it's a good idea to have the servo capstan assemblies removed and the cable clutch settings adjusted to vendor specifications. This is something that is usually ignored unless there is a problem, but by keeping the clutch settings optimal, you'll prevent a lot of problems and save money in the long run. After reinstalling the servo, recheck and adjust, if necessary, the bridle cable tensions.

Here's another important maintenance tip: A lot of owners have their airframes treated with ACF-50 corrosion-prevention fluid. The treatment consists of spraying the ACF-50 into every nook and cranny of the airframe for maximum corrosion prevention. It's been found, however, that the ACF-50 fluid can damage the material used on autopilot and trim servo clutches, so if you've having the ACF-50 treatment, make sure the facility covers your servo clutches before spraying.

Autopilot Testing

Your autopilot should have a manufacturer-recommended test procedure in the autopilot operational manual, and this is the test you should perform before any flight where you'll be relying heavily on the autopilot. You can also use this test as an initial troubleshooting test to track down where any problems are occurring, but its chief value lies in helping you find troublesome malfunctions before they crop up unexpectedly. While performing the test, check the failure monitors for each system as well. These include warning flags, or the audible tone that tells you the autopilot has been shut off.

Once you've engaged the autopilot during the testing, determine if the control wheel turns left or right with the respective autopilot commands. Check if the control wheel responds properly after receiving pitch commands from the autopilot.

Autopilot Flight Testing After New Installation

Following installation, some adjustments might need to be made to your new autopilot. These tests can reveal adjustments that need to be made or problems you might have to get fixed.

When flying straight and level on autopilot, there should be no altitude gain or loss and no turning tendency. You shouldn't notice any pitch problems at the moment you engage the autopilot. When adjusting roll trim, the distance between each wing and the horizon is used as a reference, backed up by the turn coordinator. The compass is used as a reference to adjust the rudder trim if a yaw damper is installed.

With the autopilot engaged, check again for wings level, then engage altitude hold and monitor for any noticeable bumps, porpoising, or erratic movement. Bumps could indicate that the pitch centering needs to be adjusted. If the autopilot is holding the chosen altitude, engage the heading mode and make sure the autopilot tracks the heading bug properly. Check for noticeable left or right wing dropping in tracking the heading bug, possibly an indication of the gyro needing rotational alignment.

Two other problems you might encounter with the heading bug could be caused by the heading output from the heading indicator not being within tolerance. Or, the computed output of the flight computer might not be accurate. Except for wiring problems and electromagnetic interference, those are the two most common heading-tracking problems you'll run into.

If you notice that the autopilot isn't smoothly tracking in NAV mode and following the CDI precisely, check the NAV receivers first. The tracking of any signal by the autopilot is only as good as the quality of the signal received. To monitor the NAV receivers, watch the CDIs while tracking a VOR without the autopilot. The CDI needles should be steady and shouldn't drift from left to right.

Warning: On some older King autopilots, if you test the marker beacons during an autopilot-coupled ILS approach, the autopilot will respond to the resultant gain reduction.

Troubleshooting Checklist

This section lists some symptoms and their possible causes.

Hardover (nose pops up or down suddenly):

1. Trim switch sticking.
2. Computer failure.
3. Wiring failure.

Autopilot fails to respond to aerodynamic changes:

1. Pick-offs in gyro open.
2. Open wiring between gyro and computer.
3. Defective gyro.

Altitude-hold porpoising:

Divergent, gets worse and doesn't recover:

1. Cable tension too loose.
2. Servo clutch slippage.
3. Defective servo.

Damping, eventually recovers:

1. Defective servo.
2. Insufficient current.
3. Noise injected into pitch circuitry.

Control-wheel shudder:

1. Broken wires.
2. Feedback circuitry failure (servo).
3. Computer needs adjustment.

Autopilot disengages by itself:

1. Servo clutch torque set too high.
2. Flight control or bridle cable tension set too low.
3. Defective trim switch.
4. Defective trim servo.
5. Chafing wires.
6. Defective disconnect switches.

Autopilot constantly searching while NAV tracking:

1. NAV antenna
2. receiver problems.

Autopilot drifts off heading while tracking heading bug:

1. Trim knob not centered.
2. Flaps not centered or trimmed for straight-and-level flight.

Never Say INOP

S-TEC explains why it is important to provide detailed information to technicians:

> It is critical to the efficient repair of these systems that the pilot know how to communicate with Avionics technicians. The last thing a technician wants to read on his or her work order is the

S-Tec 55 autopilot control.

squawk, "INOP." It tells the technician nothing other than there is a problem and the pilot didn't take the time to be detailed. Sure, some of the problems are intermittent and the technician doesn't want to see it on a work order, but that is the real world. Intermittents do exist and as long as they do, technicians will have to deal with them. Sure, it can be very frustrating to repeatedly to have the plane come into the shop, but don't you think the pilot is a little ticked about it also? Keeping all that in mind, the term "INOP" should be avoided in lieu of a more descriptive explanation of what happened. Avoid the word "INOP."

To say an autopilot "porpoises in altitude hold" is better than just INOP, but follow up by adding more detail like, "about ten minutes after the altitude hold is engaged it begins to oscillate up and down with minor pitch changes. They get larger as time goes on until after about twenty five minutes you have to disengage the autopilot and start over." Now that the problem is identified, along with its characteristics and function of time, the technician can begin to make an accurate diagnosis. In fact, this squawk most likely will result in a fix in a very timely fashion.

Describe all the symptoms, even if you think the symptom has nothing to do with the problem. Mentioning to the tech that a radio noise seems to have the same frequency as the rotating beacon may point to the source of the problem. If you hadn't mentioned that, it might take the tech hours to isolate the difficulty, with every hour appearing on your bill at forty dollars an hour or more.

When taking the airplane to the shop offer to talk to the tech who will do the repairs. Hearing the symptoms directly from you will help them more than second hand information. If that's not possible, the pilot should provide a phone number where the pilot can be contacted during normal working hours to get more details.

Finally, when an anomaly is first noticed, before it requires going directly to an avionics shop, make a note as to what it is, when it occurred and the initial symptoms. That, along with the subsequent history of its occurrence will be additional information of value to the tech when the system has degraded to the point it must be repaired.

Scott Howard of S-TEC has this to say about troubleshooting oscillations in the pitch or roll axis.

Often the symptoms are described as one or a combination of the following:

Roll: "The wings rock back and forth"; "Wingrock"; "The autopilot is sluggish in Roll"; "It overshoots the heading bug"; "It S-turns in Nav"; "It stairsteps in the turn"; "It's slow to respond in roll"; etc.

Pitch: "It porpoises, hunts on altitude and/or in vertical speed mode"; "It overshoots and porpoises on altitude capture"; "It has excessive loss of altitude in turns with porpoise when coming out of the turn"; etc.

The first troubleshooting step should begin with the servo involved. Check for anything that might be preventing the servo from moving the control surfaces within the time/speed parameters commanded by the autopilot electronics. This could cause the above symptoms.

Next check the physical installation of the servo. Check for loose bridle cables which would cause a delay in the control surface response as the servo pulls out the slack. *Note:* Most S-TEC roll and pitch servos require a bridle tension of 15 in. lbs. Another good thing to check is the primary cable tensions to the airframe manufacturers specifications.

If all of the above checks out properly, proceed to the servo motor and check for high starting voltage. If the motor does not respond at the proper input voltage, an error signal input will cause a greater command output to the servo. When the high starting threshold of the motor is reached, the motor will turn too fast causing over-correction thus inducing oscillation. A high starting voltage is especially critical in the lower authority Nav mode in roll-creating tracking problems. These symptoms usually show up before you see problems in heading mode.

Often the cause of this condition is wear in the D.C. motor brushes, thus creating carbon buildup on the commutator. In extreme cases, this can be detected by measuring the resistance across the motor leads (pins 3 and 4) with a digital ohmmeter. Twelve-volt motors should show a resistance of 3 to 4 ohms. Twenty four-volt motors should show a reading of 8 to 10 ohms. If you see a reading 50 percent higher than these, that would indicate a high starting voltage problem. *Note:* A normal resistance reading does not preclude a motor problem as the culprit. An ideal starting voltage would be equal to or less than 1.5 volts for a twelve-volt motor, and equal to or less than 2 volts for a twenty four-volt motor.

In some cases these symptoms can be improved or eliminated by performing a servo brush seating procedure. This can be done by applying full aircraft voltage to the motor leads and allowing the motor to run for four minutes in both directions. Since the motors seldom see full rated voltage, the increased speed will often break the carbon glaze free from the commutator and the brushes will seat uniformly. This procedure is most easily accomplished through the computer connection utilizing the S-TEC 9524 break-out box. Jumper the motor leads to A+ and ground leaving the AP computer unplugged (AP master switch on).

If symptoms persist the servo must be removed for cleaning and service of the motor.

Fly It Better

Jim Irwin, Vice President of Products for S-TEC, explains how to more effectively fly the autopilot under IFR:

Airplanes maneuver around the three axes of roll, pitch, and yaw. During autopilot-controlled maneuvers, the commands are typically limited to roll and pitch with yaw commands being a reaction to yaw upsets requiring damping, normally through a yaw damper.

Commands through the roll and pitch axis are made to accomplish a specific objective, i.e., to turn or not turn and to climb or descend or to hold an altitude.

In instrument flight, maneuvering standards are taught. A standard rate turn, a specific bank angle for normal maneuvering, a specific intercept angle for certain course offsets, a 500 FPM climb or descent for short climbs or descents or when nearing the desired altitude. These standards create a maneuvering, rate, and position judgment capability the pilot uses to unconsciously judge where he or she is, or where he or she will be. This is based upon the pilot's intuitive judgment of the elapsed time and the rate of movement provided by these standards. This, in turn, provides the pilot with a degree of instrument-scan flexibility; if the pilot has 1000 ft to lose at 500 FPM, he or she has two minutes to use to concentrate on some other requirement.

This use of time and rate is one of the techniques available in instrument flying to increase precision and situational awareness and allow time for housekeeping.

S-TEC rate autopilots lend themselves to this capability, not only by handling many of the flying duties, but also by providing predictable maneuvering rates in both turn and climb or descent maneuvering.

To maneuver in the pitch axis, S-TEC full function autopilots use vertical speed (Vs) as the command reference. Our autopilots use either a time based modifier switch that modifies the Vs reference at a specific rate (Systems 60-2, 60PSS, and 65) or a digital Vs selector (System 55 and Altitude Selector/Alerter).

New owners and operators of S-TEC System 60-2, 60PSS, or 65 Autopilots often hold the modifier switch for up or down until the aircraft attitude has changed to the attitude that they believe will produce the desired climb or descent. However, since the device is time-related, this procedure will produce a larger attitude change than desired. The correct method is much faster, easier, and less obtrusive, especially during instrument maneuvering flight when you must divide your attention between so many activities. The up or down modifiers revise the reference Vs at the rate of 160 FPM per second of activation. Therefore, to initiate a standard rate descent or climb (500 FPM) it is only necessary to select Vs mode and then push the desired modifier and hold it for about three seconds. During this period the aircraft will begin a very smooth pitch change in the direction commanded, verifying the command. At the end of the three seconds' input, the Vs will be at approximately one-half the desired rate. Releasing the modifier button the pilot can resume the instrument scan or direct his attention to other duties. Thus, a turn or intercept can be accomplished while the autopilot is completing the Vs command. On the next scan cycle the pilot can verify proper operation and see how close the command came to 500 FPM.

For a shallow descent, such as that from a procedure turn altitude to a lower intermediate segment altitude, a 200 FPM descent rate requires only slightly more than a one second activation before you can return to your scan and other approach chores. Push the modifier for the required period, then get on with other duty items while the maneuver is initiated and developed.

For significant changes in vertical speed, such as 1000 FPM or more, input a three-second, 500 FPM command to get things started, then go to the next step in the maneuver such as initiating a turn. Finally, return to add the additional Vs command. This management technique allows you to get the course and altitude changes started immediately, which is often appreciated by ATC, while not spending an excessive amount of time on any one chore.

While proper use of the time-based modifiers is both easy and fast, the use of the optional S-TEC Altitude Selector/Alerter is even faster, particularly if you need a large Vs change. With this system installed, the pilot simply dials in the desired up or down Vs on the selector. Next, he selects the autopilot Vs mode, observes the initial pitch over, and then goes on with the instrument scan. Returning to verify the desired result on the next scan is the last step. To increase or decrease the original Vs selection, it is only necessary to rotate the Vs selector knob to the new Vs command value. If you have chosen, or been assigned an altitude, you can input that altitude to automatically level off there. An altitude-alert chime advises you of the approaching altitude. Beginning to make IFR maneuvering pretty easy, isn't it?

The S-TEC System 55 has a digital vertical speed selector built in. As with the optional Selector/Alerter added to Systems 60-2, 60PSS, or 65, you can input the desired Vs and then select the autopilot Vs mode, observe the initial pitch change and resume your scanning or housekeeping. Once again, the last step is returning on the next scan or two to verify the vertical speed.

There are times, in IMC, when you must maneuver the aircraft in response to ATC commands while simultaneously familiarizing yourself with new and unexpected approach or navigation problems. It is during these busy times in the cockpit that understanding the maneuvering rate and time relationship will be very helpful. You won't have to feel panicked by changes and clearance amendments—simply dial in the initial requirements, heading, and Vs, and use the maneuvering period to catch up with your other duties. Try it, it works.

GPSS: Details, Details, Details

Jim Irwin, Vice President of Products for S-TEC, writes,

We have received a significant number of inquiries, questions and comments concerning GPSS since we introduced the new autopilot function at AirVenture '99—Oshkosh.

The following will attempt to explain the development of the steering signal to the autopilot's use of that signal. The information is general in nature and not a description of the processes used by any specific navigator. In addition, the information is being provided by a pilot, for pilots who want a little better understanding of the overall process.

Steering signals or commands produced by some long-range navigation systems and radio couplers used in autopilots basically do the same thing. Why then, are steering commands typically more accurate and authoritative while also being smoother and more decisive?

This question gets right to the heart of things. The difference, like so often, is in the scope and quality of the data available. The GPS navigator has more data available, and the quality of that data is superior to data supplied by other navigation systems.

The GPS navigator has access to (often updated more than once each second) the aircraft's position, direction of movement, groundspeed, desired course, actual course, and cross-track error (amount off-course). From these elements, it can determine whether it is closing on the desired course and at what rate.

The autopilot's radio coupler on the other hand only has available radio position, the rate that the aircraft is closing on the course, and the course error (the difference between the course and the heading of the aircraft). That's it. The quality of the data available to the coupler is often, in the case of VOR, adversely influenced by factors such as radio noise, antenna masking, terrain interference, signal strength, and phase shift variations. The coupler also has to struggle with behavioral anomalies caused by geometry and distance, since VOR is associated with angles. The course error in distance varies with distance from the station at the same radio error amplitude. That is, two-needle widths' course error is not much error at two miles, but it may take five minutes to resolve at sixty miles.

Some of the behavioral anomalies still exist even when the coupler is used with GPS L-R error commands used in standard GPS tracking through a CDI. In this case, the courses do not converge on the station, as with VOR, but are instead constant course-width signals.

This characteristic of GPS aids the intercept and tracking maneuver by providing consistency. However, the radio coupler still does not process the aircraft's actual direction over the ground or its groundspeed, and therefore, the coupler cannot factor this vital information into the navigation equation. Additionally, radio couplers designed for VOR tracking include noise filters to reduce the effect of VOR signal noise. This filtering is detrimental to GPS L-R tracking because the GPS signals are NOT inherently noisy, and the filter's time constant induces lag, which causes less precise tracking of the GPS signal. The filter helps VOR tracking but is detrimental to GPS tracking. For these reasons and more, autopilot radio couplers are full of compromises that affect tracking performance with VOR or GPS navigation sources in specific navigation situations but not in others.

If GPS has more and better data available, how does the navigator compute it? For the most part, the roll steering functional block diagram describes an algorithm that is almost identical to a normal autopilot radio coupler, except that it knows the actual ground track. One of the primary steering ingredients is the cross track deviation, or the selected cross-track summed with cross-track distance. This signal, which represents how far off course we are, is fed to a limiter that is used to limit the angle of the intercept. Let's call this signal #1.

The other principle ingredient is the rate component that is derived by summing the desired track and the actual track signals. This signal, which is track angle error, is fed to a multiplier that multiplies the track angle error with ground speed. Let's call this signal #2. Signal #2, track angle error, multiplied by the ground speed yields the rate of closure to the desired track.

At this point, we have developed a signal representing our cross-track deviation and provided it a limiter so it can intercept at a reasonable angle.

We have also developed a signal representing our rate of closure to or departure from the desired track.

The next step is to sum signals #1 and #2 such that the cross-track deviation commands a rate of closure. The result is then fed through a roll-angle (bank) limiter and a roll-rate limiter so that we don't over-command either our rate of roll or the desired bank angle. This is then sent to the autopilot.

The two most common forms for the steering outputs to the autopilot are an analog AC signal, or a digital output in the popular ARINC 429 data bus format. If the output is analog and can be matched to the heading error signal, it is often only necessary to intercept the HDG Bug input to the autopilot with a switch that allows exchanging HDG with the steering signal. The autopilot then flies the steering command as it would HDG error.

The more popular steering signal is the 429 format. The navigators that output the 429 signal are all digital, meaning the data elements are already available in digital format. In our case, we take the 429 signal and convert it to the equivalent of a HDG signal. However, we provide it with more turn rate capability than we do the HDG channel. This is to assure that the autopilot will be able to maneuver adequately when conducting one of the standardized GPS approaches that have 90-degree turns from the initial approach segment to the intermediate segment. This is particularly important at approach speeds.

For the pilot, life is simple. Select the desired way-point on the navigator, and select GPSS mode on the autopilot or on the HDG/GPSS selector switch, and watch modern science in action.

Now that we have the interface to the autopilot covered, we should discuss the navigator again for a few more moments. Since the system provides the ground-speed component along with the position and direction of flight, we now have a real-time velocity vector representing aircraft direction and ground speed that is extremely accurate. The system knows where it is relative to the desired track; it knows the rate at which it is closing on the track, and it knows the actual ground-intercept angle and ground speed.

The system can now plan turns in terms of turn radius over the ground, as opposed to simply banking to make a turn radius through the air. Since it knows the ground speed of the airplane, the system can project where the intercept turn should start in order to join the course without over/under shooting the sought after asymptotic course capture. The navigator is programmed for a specific bank angle, and it computes the turn initiation point based upon a turn at the programmed angle. The systems typically include a bank limit and often a bank rate limit, as indicated above.

Bank limits vary with different systems but the steering formulas for transport category aircraft often use 25 or 27 degrees at low altitudes, decreasing above about 20,000 feet linearly to 15 degrees of bank at 35,000 feet and above. Often "direct to" commands will use a little higher bank angle limit for better response in that maneuvering situation. Helicopters and light GA aircraft are often provided with lower turn limits, often 15 to 20 degrees of bank. Roll rate is often limited to 3 degrees/second.

As the system commands maneuvers to join the desired course, it tracks the actual ground vector being produced by the moving airplane and adjusts the turn to assure that the new course will be joined accurately.

Steering commands to join a course do not have the indecisive quality of many radio coupler turns, that is, no bobbing and weaving as the navigation needle wavers from interference, antenna blanking, etc. Typically the aircraft flies straight and level, with little or no turning motion, until such time that a turn initiation point is reached. Here a smooth turn is initiated at a comfortable turn rate until the correct bank angle is reached and held until the roll out to on-course. Cross-wind correction angles are established quickly and precisely and with little maneuvering. Think about it. The navigator knows when the aircraft is off the desired track by only a few feet and can, therefore, often make a correction before there is a CDI movement.

I hope the additional detail is valuable and that you have found information here that you have not seen before. Someday, all aviation GPS Navigators will output steering commands, and all autopilots will have dedicated steering modes like GPSS. Too many technical and performance differences exist between VOR and GPS navigation systems for any existing radio coupler to be able to do a perfect job on both. A perfect job takes a dedicated interface like GPSS by S-TEC.

Special thanks to Allen Grime and his crew at Century Flight Systems, who did their best under difficult circumstances to confirm accuracy of the information relating to their Century 41 autopilot.

Also, thanks to Jim Irwin, Vice President of Products at S-TEC and Scott Howard of S-TEC for their assistance and material for this book. It should be noted that S-TEC is now owned by Meggitt Avionics.

Autopilot Manufacturers

The following autopilot manufacturers' addresses can be found in the Sources section of this book.

Bendix/King	Collins Avionics
Brittain Industries	Honeywell
Century Flight Systems	S-TEC
Sigma-Tek	

CHAPTER 16

Transponders and Collision Avoidance Systems

Transponders are like the DME in that they also are pulse-type avionics. Whenever a transponder is interrogated by a radar system, the transponder "replies" with a pulse of information containing the four digits that you have dialed in to the face of the transponder and, if equipped with an encoder, your altitude.

During the Second World War, the British developed a top-secret radar transceiver. Carried aloft by allied planes, it was designed to respond to radar interrogation signals with a coded transmission. This code would allow the allied forces at the land based radar station to identify if it was a British or German aircraft on their radar screens. Interestingly, in interests of secrecy, the transceiver also contained a bomb that would explode should the plane crash (glad they don't put them into transponders today). This was to be the familiar IFF, for Identification Friend or Foe system used in today's military aircraft. This device metamorphosed into the transponder we all love and cherish.

All transponders operate on 1090 MHz. For Mode A, the transponder gives back only the four-digit code. For Mode C includes the

KT 76 panel-hounted transponder.

altitude. This is done 600 times a second but only 20 to 30 responses occur during the radar beam passage. If two radar interrogations occur nearly simultaneously, the transponder response may become garbled at the radar site. This often leads air traffic control (ATC) to claim that your transponder isn't working properly. Your recycling the transponder is a way of changing the response sequence. If one radar location had no difficulty with your transponder refer the problem site to them. Often different locations are using widely age-different systems. Ask that the radar tapes be saved so that the FAA may analyze of the problem.

The following applies to all non-turbojet planes: An airplane may be without an altitude encoder and operate only on MODE A or ON. Under Mode A, ATC will expect you to maintain either an assigned or agreed upon altitude and to report changes. Flight with Mode A is somewhat restricted. If you know that your transponder does not have Mode C capability, be sure to advise ATC. Know the following restrictions.

Aircraft above 10,000 feet are required to have an operative transponder with Mode C. Aircraft in Class C airspace or above the outer perimeter of the Class C airspace up to 10,000 feet are required to have a transponder Mode C. Any flight above the Sacramento Class C comes under this last requirement. Any aircraft in a Class B or operating within thirty nautical miles of the Class B primary airport is required to have an encoding transponder. Exceptions are made for aircraft without electrical systems and high mountain flights within 2,500 feet of the surface. The transponder and encoder system must be inspected every 24 months.

An aircraft without an operating transponder shows, if at all, as a primary target. All transponder targets are called secondary. Under MODE C, or ALT, your aircraft will have an encoder, which tells ATC your altitude. ATC will always need to know if your altitude encoder is operating correctly. This altitude encoder is crosschecked by ATC with your altimeter setting via radio. They will remind you of the current altimeter setting and perhaps ask you to switch to MODE A if your encoder is off by more than 300 feet. An error of 300 feet makes the transponder unsafe to use for traffic avoidance purposes.

If you should experience a transponder failure, be cautious about accepting flight into a radar environment where radar is the prime system. Once you land at a Class C airport without a transponder, you may be unable to get out.

On the ground, the transponder should be set to standby and the codes switched to 1200. The codes used to land are no longer valid; they are simply assigned by ATC to someone else. Going to standby stops the squawk but allows the transponder to stay warm and ready for operation when needed. The start takeoff, emergency, and post-landing checklist should have the transponder as a checklist item. Whenever changing codes on the transponder recommendation is that "standby" be selected during the change since it prevents inadvertent discrete codes being sent to ATC. The transponder should be turned on as you taxi onto the runway for takeoff. Use of MODE C is now required in many cases as noted earlier.

Transponder Modes

Mode A refers to the coded reply, the four digits that show radar controllers your position. Mode C shows not only the coded reply, but your altitude, too, helping controllers keep traffic in busy areas separated. Neither Mode A nor C is selective enough to provide sufficiently accurate information in busy airspace and can even cause erroneous display of near-midair collisions that never even happened.

Ground radars constantly sweep the skies, interrogating all airplanes within range. If the radar is sweeping a congested hub, such as Los Angeles International Airport and the airspace surrounding it, an enormous number of interrogations and replies are taking place. Because of the huge amount of traffic and short duration of time between the transmissions, transponder replies can overlap each other and confuse the ground radar. When flying close to radar sites, pilots of interrogated aircraft might have to set their transponder sensitivity to the "low" position to prevent false targets. Some pilots might forget to switch the transponder back to "high" and might get a call from ATC gently hinting for them to "please recycle your transponder."

In the past few years, transponders have become increasingly important, especially after some rather unfortunate midair collisions occurred. The FAA now requires that pilots of airplanes that are transponder-equipped turn them on when flying in controlled airspace. Since December 30, 1993, all aircraft flying within thirty nautical miles of the center airport of Class B airspace (TCA) must be equipped with Mode C capability to transmit altitude information to ground controllers along with the four-digit transponder code.

There is an ongoing regulatory attempt to mandate Mode S transponders. For Part 91 operators, the latest rule specified that after a certain date, only Mode S transponders could be installed. That rule is now on hold and, at least as of this writing, Mode C transponders were still being manufactured and installed. Actually, brand new, solid state Mode C (under TSO-C74b or TSO-C74c) units are being released the first part of 2001. Per Code of Federal Regulations Title 14ATC 91.315, transponder equipment installed must meet the performance and environmental requirements of any class of TSO-C74b (Mode A) or any class of TSO-C74c (Mode A with altitude reporting capability) as appropriate, or the appropriate class of TSO-C112 (Mode S).

If the time comes when you need a new transponder and the new rule takes effect again, you might want to hurry up and buy a Mode C transponder before they're outlawed. They will most likely grandfather the existing Mode C version, phasing in the Mode S as the Mode C units fail. Mode S transponders are very expensive, and it doesn't look like the price will ever be as low as Mode C.

One important note about Mode S transponders, when and if they ever become required: The transponders rely on a discrete code that is assigned exclusively to a particular airplane, like your Social Security number. So if you were having a Mode S transponder installed you would have to make sure you were assigned a discrete code for your transponder at the same time. The FAA is concerned that Mode S transponders are being installed without discrete codes being assigned or using generic codes like all ones or zeros. This would prevent them from specifically tracking a plane at any given time at any given place.

Encoders

There are two ways to provide altitude information along with your Mode A transponder code. The least expensive is to add a blind encoder or digitizer, which is a small aluminum box that can fit easily under the instrument panel. If you don't already have an altimeter or you'd like to have a spare altimeter, you can go the more expensive route of getting an encoding altimeter. This fits into a standard 3-inch hole in your instrument panel.

In either case, once you have a blind encoder or an encoding altimeter installed, the Mode C or "alt" setting on your transponder will allow you to fly within thirty miles of Class A airspace and into airports inside Class A airspace. Because the encoder installation requires the installer to open the static line to "Tee" in the plumbing from the encoder to the static system, a static test must be conducted per Advisory Circular 43-6A. The encoder must be calibrated to comply

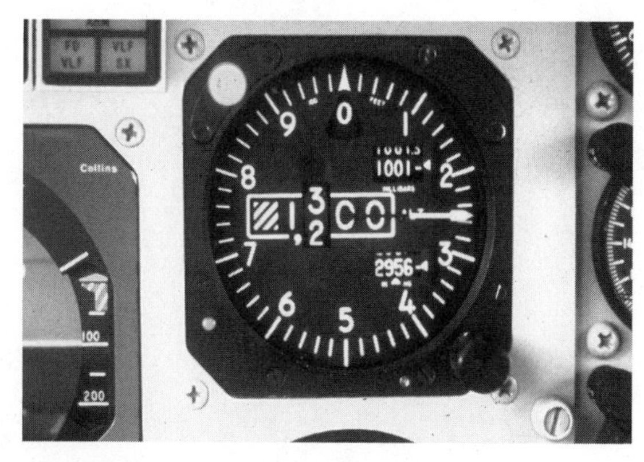

Encoding altimeter mounted in the pilot's panel.

ACK remote encoder, requires only power, wiring to transponder, and a "T" fitting into the static lines. This, of course, requires the static system be recertified per Title 14 of the Code of Federal Regulations (14 CFR) part 91, section 91.411, guided by AC 43.13-1b, Appendix E.

with 91.36 and 91.172, which is outlined in 43-6A.

Transponder Installation Tips

Because both DME and transponders are pulse-type equipment, the transponder can have a side effect of suppressing the DME signal (see Chapter 12). Some avionics shops and factory installers wire the DME-suppression protection into the system but don't hook it up. This leaves the owner the option of having it hooked up later if DME suppression becomes a problem. It's never a good idea to hook it up if there isn't a problem, however, because the feature can also actually mess up DME operation.

The panel-mounted transponder is recommended because the remote installation means there's a lot of wiring from the instrument panel to the black box. If it's all the way in the back of the airplane, this means removing a bunch of upholstery and possibly introducing many areas of potential errors. One major hazard is piercing avionics wiring with upholstery screws when reinstalling upholstery.

Transponders operate at a frequency that isn't susceptible to the same interference problems that bedevil communications, NAV,

Narco 150R transponder.

ADF, and LORAN systems. To ensure no interference from other noise generators and precipitation static, however, it's important to install a properly bonded antenna on a 6- to 8-inch ground plane, free of interfering obstructions, such as gear doors or other antennas.

The transponder antenna is the short little blade type, short, stubby tube-style, or the cheaper rod with a little ball on the end. As you might remember from Chapter 12, the rod-type antenna is very sensitive to the long arm of the airplane washer and can be easily bent or damaged. I've seen cases where a mechanic who broke off the antenna simply stuck the rod back into the base, hoping no one would notice that it was broken. Amazingly, the transponder actually worked, until the rod fell out. Stick with the blade antenna; it will last longer and provide better service.

Transponder Testing

There are three tests done on transponder systems:

1. Self-test.
2. FAA-mandated ramp test.
3. Flight test.

First, most transponders have a self-test function you can perform. On a King transponder, for instance, you turn the knob all the way to the right temporarily, then let the knob go. This causes the transponder to generate a simulated signal internally, lighting the reply light. This isn't much of a test, so don't absolutely rely on it to tell you if the transponder is really doing its job.

The second test is the FAA-mandated ramp test, required under FAR 91.413.

The final test is the flight check. This should be performed to evaluate overall operation and compatibility with other aircraft systems if this is a new installation. The transponder should provide a good signal at 6,000 to 12,000 feet at a range of 50 nautical miles from the ground radar, Your encoding altimeter or blind encoder should give your altitude plus or minus 125 feet at all times. Flying

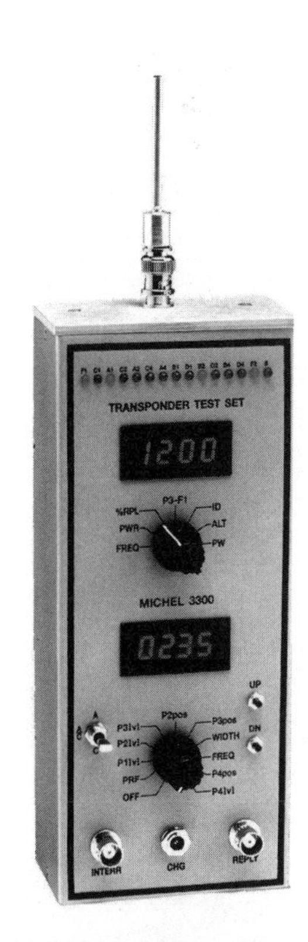

Hand-held, inexpensive Michel 3300 transponder ramp test set. Very easy to use and allows for complete certification of transponder/encoding system.

in a circle, the controller's radar should not take more than two sweeps without seeing your transponder. Finally, you can fly a simulated surveillance approach down to 200 feet. The radar should not miss or drop your target during any given 10 sweeps during your low pass at 200 feet. These distances and altitudes are approximate because of the different power outputs of each manufacturer's transponder. Use these only as a guide to determine if your transponder is meeting minimum conditions.

Altimeter System and Altitude Reporting Equipment Tests and Inspections

Title 14 of the Code of Federal Regulations (14 CFR) part 91, section 91.411 clearly states that the altitude reporting system be tested every 24 months.

(a) No person may operate an airplane, or helicopter, in controlled airspace under IFR unless—

  (1) Within the preceding 24 calendar months, each static pressure system, each altimeter instrument, and each automatic pressure altitude reporting system has been tested and inspected and found to comply with appendix E of part 43 of this chapter;

  (2) Except for the use of system drain and alternate static pressure valves, following any opening and closing of the static pressure system, that system has been tested and inspected and found to comply with paragraph (a), appendices E and F, of part 43 of this chapter; and

  (3) Following installation or maintenance on the automatic pressure altitude reporting system of the ATC transponder where data correspondence error could be introduced, the integrated system has been tested, inspected, and found to comply with paragraph (c), appendix E, of part 43 of this chapter.

(b) The tests required by paragraph (a) of this section must be conducted by—

  (1) The manufacturer of the airplane, or helicopter, on which the tests and inspections are to be performed;

(2) A certificated repair station properly equipped to perform those functions and holding—
   (i) An instrument rating, Class I;
   (ii) A limited instrument rating appropriate to the make and model of appliance to be tested;
   (iii) A limited rating appropriate to the test to be performed;
   (iv) An airframe rating appropriate to the airplane, or helicopter, to be tested; or
   (v) A limited rating for a manufacturer issued for the appliance in accordance with Sec. 145.101(b)(4) of this chapter; or

(3) A certificated mechanic with an airframe rating (static pressure system tests and inspections only).

(c) Altimeter and altitude reporting equipment approved under Technical Standard Orders are considered to be tested and inspected as of the date of their manufacture.

(d) No person may operate an airplane, or helicopter, in controlled airspace under IFR at an altitude above the maximum altitude at which all altimeters and the automatic altitude reporting system of that airplane, or helicopter, have been tested.

Altimeter Testing Precautions

Altimeter tests are performed in accordance with FAR Part 43, Appendix E. If the altimeter test is to be performed with the instrument installed in the airplane, the following guidelines should be observed:

1. The static leak test should be conducted first to assure that there are no static system leaks that might affect altimeter readings.
2. Do not test directly after a flight; allow the altimeter to stablize.
3. Test altimeter separate from aircraft system to ensure it is functional.
4. If vibration is applied to the instrument, make sure it is not conducive to concealing a defective altimeter.

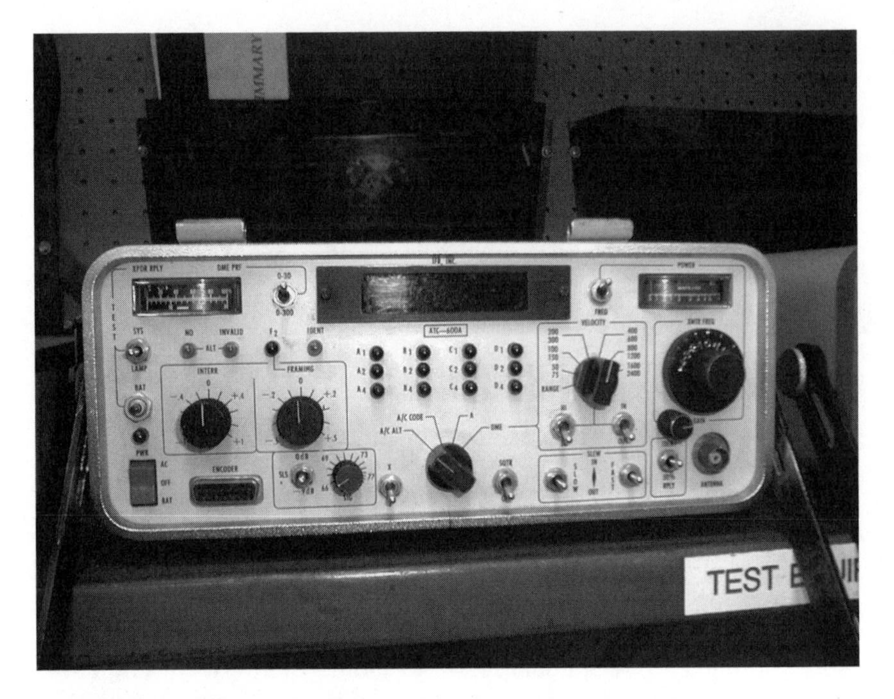

IFR 600A Transponder/DME ramp and bench test system.

ATC Transponder Tests and Inspections

Title 14 of the Code of Federal Regulations (14 CFR) part 91, section 91.413 states:

> (a) No persons may use an ATC transponder that is specified in 91.215(a), 121.345(c), 127.123(b), or Sec. 135.143(c) of this chapter unless, within the preceding 24 calendar months, the ATC transponder has been tested and inspected and found to comply with appendix F of part 43 of this chapter; and
>
> (b) Following any installation or maintenance on an ATC transponder where data correspondence error could be introduced, the integrated system has been tested, inspected, and found to comply with paragraph (c), appendix E, of part 43 of this chapter.
>
> (c) The tests and inspections specified in this section must be conducted by—

(1) A certificated repair station properly equipped to perform those functions and holding—
   (i) A radio rating, Class III;
   (ii) A limited radio rating appropriate to the make and model transponder to be tested;
   (iii) A limited rating appropriate to the test to be performed;
   (iv) A limited rating for a manufacturer issued for the transponder in accordance with Sec. 145.101(b)(4) of this chapter; or
(2) A holder of a continuous airworthiness maintenance program as provided in part 121, 127 or Sec. 135.411(a)(2) of this chapter; or
(3) The manufacturer of the aircraft on which the transponder to be tested is installed, if the transponder was installed by that manufacturer.

Of course, you can't perform a calibrated transponder or static check if the altimeter isn't accurate. Title 14 of the Code of Federal Regulations (14 CFR) part 91, section 91.411 covers the calibration of the altimeter and encoder.

Pitot Static System Check

Title 14 of the Code of Federal Regulations (14 CFR) part 91, section 91.411 also covers the certification of the altimeter and static system. Also, per this regulation the transponder and encoding system cannot be certified as being accurate if the static pitot system isn't properly checked. The same is true for the altimeter. As part of performing this test, the altimeter, transponder and encoder can also be checked and certified under Title 14 of the Code of Federal Regulation (14 CFR) part 91, section 91.413, which covers the transponder testing and certification. Many shops routinely perform the reporting system checks at the same time the static system is evaluated. The following is a step-by-step procedure to check to confirm the plumbing is intact, without leaks.

Pitot Static System Test Requirements

No matter what type of aircraft you maintain, at some point you will need to perform a pitot or static leak check. If you maintain an aircraft that operates in IFR or VFR class B airspace (Canada only) the static system inspection and test is required every twenty-four months. On all aircraft, anytime the static or pitot system integrity is disturbed, either by opening a line (such as installing an encoder) or removing an instrument, the system will have to be checked to ensure no leaks are present. Normally, no check is required if a quick-disconnect or self-sealing drain valve is opened, however, many times those fixtures have failed after operation and began leaking. This usually happens after a long period of not being used, and they stick in the semi-open position.

The 24-month requirements are found under FAR Appendix E to part 43 for U.S. operations; CAR 571 Appendix B for Canadian operations. A summary of the static system checks are:

1. No moisture or restrictions in system.
2. Static port heater, if installed, is operative.
3. No alteration or deformity near the static ports that could disturb the airflow.
4. Leak rate within tolerance.

Consult the current regulations for more detail and current requirements. Note that no mention is made of the pitot system. This is not a pitot/static test; the only requirement every 24 months is to test the static system, however, the pitot is tested anyhow as part of the procedure.

Test Equipment Requirements

Equipment, materials, and required tests for test equipment are specified in Section 145.47 of the FAR. Persons authorized to perform altimeter and static systems tests are identified in section 91.170 of the FAR.

The following test equipment is acceptable for testing altimeters:

(a) Mercurial barometers with accuracies specified in and maintained in accordance with Advisory Circular 43-2A.

(b) High accuracy portable test equipment (with appropriate correction card) maintained in accordance with FAR 145.47(b). It has been found that calibration checks of the test equipment in accordance with the following schedule provides a satisfactory level of performance:

  (1) Each thirty days, after initial calibration, the equipment should be checked for accuracy against:

    (i) A barometer described in (a) above, or,

    (ii) An altimeter (with appropriate correction card) which has been calibrated, within the past thirty days, against a barometer described in 7a. "The exception would be a digital altimeter which is not required to be calibrated for a year."

  (2) Each day the equipment should be checked for accuracy at station pressure using an aneroid or mercurial barometer. If the equipment is not used daily, a "before use" test may be substituted for the daily test.

There are two leak rates, depending on whether the aircraft is pressurized or not. For a pressurized aircraft, the system must be evacuated to the maximum certified cabin pressure differential. This is normally lower than the certified maximum ceiling for the aircraft. The aircraft is held at the test altitude and the leak rate is measured over one minute. The allowed leak rate is 2 percent of attained altitude or 100 feet, whichever is greater. For example, at 35,000 feet 2 percent is 700 feet. As long as the system doesn't leak more than 700 feet per minute, the system is serviceable. For an unpressurized aircraft, evacuate the static system to 1,000 feet above the ambient altitude. Measure the leak over one minute. The allowed leak is 100 feet.

These are the rules, but how does this work out in practice? There are a couple of things to take into account. Remember that one instrument and one system type receives both pitot and static pressure. That instrument is the airspeed indicator and the system is an ADC (air data computer). All aircraft have airspeed indicators, some have ADCs. It is extremely important to know if the static system being checked has airspeed or an ADC hooked up to it. Why? Because as the vacuum is increased the airspeed increases as well. To better understand what this means, take a look at the internal layout of the airspeed indicator; the airspeed needle is connected to bellows. Pitot pressure is fed into the bellows; static pressure is fed to the case and surrounds the bellows. As the air pressure in the static system decreases, the differential pressure between the ambient air pressure in the bellows and the surrounding pressure changes. For all practical purposes, to the airspeed indicator it's the same as an increase in pitot pressure. The bellows expand and the needle shows an increase in airspeed. On a non-pressurized aircraft, your test altitude of 1000 feet will equal approximately 140 knots. If your airspeed indicator stops at less than 140 knots, and you perform the test, the airspeed indicator will be ruined. If you need to perform the check on a pressurized aircraft, you will run out of airspeed long before you run out of altitude. To overcome this problem, the static system needs to be paralleled to the pitot system. This eliminates the pressure differential between the pitot bellows and the static in the airspeed case, and allowing higher test altitudes.

Nonpressurized Aircraft Static Check

Check the airspeed indicator. Should it indicate 150 knots or more, don't parallel to the pitot system. After checking for moisture, heater operation, and structural deformities, connect the test set to the external static port. Suction-cup-style adapters are available from distributors for this operation.

Test Procedure

1. Connect test set to the external static port using an adapter.*

*If an adapter isn't readily available, a suitable, temporary fixture can be made from coax cable seal with an appropriately sized tube that will attach to the test hose. The black, putty-like sealer is formed into a donut-like shape and placed around the tube. Next, place this arrangement with the tube centered over the static ports and press the coax seal tight against the fuselage. This provides a temporary, airtight seal for the duration of the test.

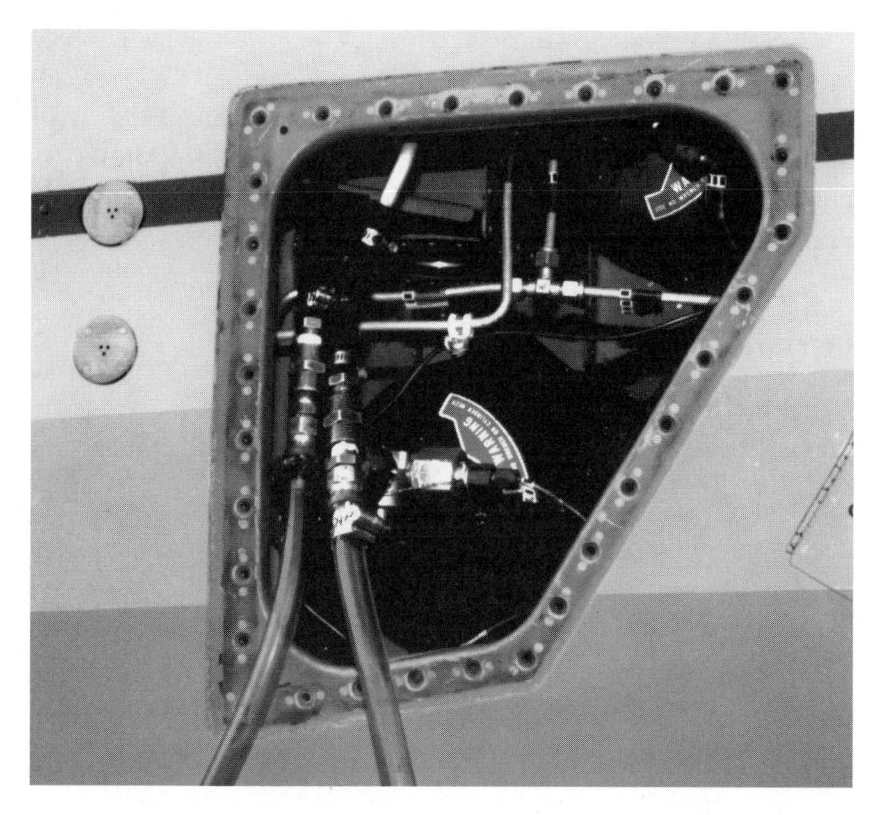

Static test connections not recommended. They should be externally connected with an adapter.

2. Tape off any other static ports with black electrical tape. Leave 3 to 4 feet of loose tape trailing from the port so it's not forgotten. Don't use masking or clear tape. Masking tape doesn't seal very well, and clear tape will be overlooked. For every person who says, "I have a good memory, I won't forget," there are two who will forget and forget they do.
3. Bring the system to 1,000 feet above what was indicated on the altimeter at ambient pressure. Don't exceed the rate limit on the aircraft VSI (vertical speed indicator).
4. Hold for 1 minute, and observe for a less than 100-feet leak rate.

If a leak is indicated, start troubleshooting by investigating the static ports and drains. If they are found to be okay, and the leaks are still present, isolate the system by splitting it in half and begin troubleshooting each half separately. When complete, come down in altitude and remove the test gear. Remember to take the tape off the static port and clean up by removing all miscellaneous tools and trash.

Pressurized Aircraft

The same procedure used for non-pressurized aircraft applies to the pressurized test, with some exceptions.

1. Do not perform the static check without some sort of professional static adapter. Having the static system test hose come off at 35,000 feet will be a suicide dive for the indicators and make the local instrument shop happy, but not your customer.
2. Limit the number of people working around the aircraft during a check. People stepping on or inadvertently pulling on hoses can be disastrous. It may be necessary to coordinate it with your supervisor. If management feels that those people must be there because of scheduling, explain how an accidental opening or clamping of the hoses can cause expensive damage. Of course, the accident waiting to happen may still happen, but now the responsibility rests with management.
3. If the test is paralleled to the pitot system, conduct a double-check and ensure the airspeed doesn't move when vacuum is first applied.

Other than removal of instruments or breaking of lines, there is no calendar requirement to check the pitot system. Leak rates are different for different aircraft, a quick check of some aircraft maintenance manuals allows from 5 knots per minute, to no leak at all. If no tolerance is given, try using 100 knots indicated and no more than 2 knots leak over 1 minute. On smaller aircraft, there is a drain at the aft end of the Pitot tube. If you don't block this drain, the system will show a leak. Covering the hole with black electrical tape is all that's required.

In summary, remember that the pitot and static system must be tested as it is would be during flight. Although frequently done, it is not acceptable to open a line, or drain, and tee into the system for the check. The test must be carried out with the system airtight. How do you know the drain cap or line you disturbed doesn't leak when you replace it? The easiest way to do this is at the static port or Pitot tube. If doing a static check, watch out for airspeed indicators and the ADCs, and don't forget to make sure you parallel the systems if you are going to high altitudes.

Disclaimer: None of the regulations or advisory circulars referenced in this chapter should be used as final. They were written only to clarify and simplify the procedures required as a part of the certification process for transponders, altimeters, and encoders.

Collision-Avoidance Systems

Most manufacturers are making collision-avoidance systems compatible with the TCAS II (traffic collision alert system) standard. TCAS-equipped aircraft provide pilots with traffic advisories and resolution advisories (advice on maneuvers to avoid conflicts) on "threat" aircraft that might come too close or pose a midair collision threat. TCAS gets its information from threat-aircraft transponders. TCAS systems are very expensive and must include costly Mode S transponders, with some of the lower-cost systems starting at over $40,000. TCAS I equipment, which gives only traffic advisories, but not resolution advisories, is mandated on aircraft with 10 to 30 seats. Mode S transponders are not needed in aircraft with TCAS I installed.

Ryan International, a company owned by Paul Ryan (who designed and marketed the Stormscope® before selling that company), designed the lower-cost TCAD (traffic collision alert device) system for light aircraft owners who want some measure of additional protection from crowded skies. TCAD also perceives threat aircraft via their transponder outputs, displaying threat aircraft range, but not threat aircraft-bearing information.

A company called Flight Technology International offers the Airtrac moving map/collision-avoidance system that displays threat aircraft that have the company's Airtrac system installed.

Transponder Manufacturers

These transponder manufacturers' addresses are listed in the Sources section at the end of this book.

Aire-Sciences
Becker Avionics
Bendix/King
Collins Avionics
Narco Avionics
S-TEC
Sigma-Tek
Teffa Avionics

Encoder Manufacturers (Altimeters and/or Blind Encoders/Digitizers)

These manufacturers produce encoding devices; see Sources for addresses.

ACK Technology
Aero, Mechanism
Aerosonic
Ameriking
Bendix/King
Kollsman
Narco Avionics
Pointer
Trans-Cal Industries
United Instrument

Collosion-Avoidance System Manufacturers

These companies offer collision-avoidance systems; see Sources for addresses.

Bendix/King
B.F. Goodrich Flight Systems
Collins Avionics
Flight Technology International
Honeywell
Ryan International

CHAPTER 17

Emergency Locator Transmitter (ELT)

Emergency locator transmitters are simply transmitters designed to activate when subjected to a specified G-force load upon making hard contact with the earth (see vendor specifications for details). ELTs transmit on 121.5 MHz for civilian reception and 243.0 MHz for military reception. There are some new regulations for ELT design that you should know about, and some features that new units offer that you might want to consider if you are buying a new ELT.

ELTs frequently are forgotten or ignored because they don't do anything until a crash landing or if someone accidentally activated one to irritate the FAA. Unfortunately, while ELTs have saved many lives, they also have caused a lot of aggravation, false alarms, and unnecessary expenditure of search-and-rescue effort. In 1991, nearly 97 percent of all ELT signals were reportedly false alarms. Furthermore, a 1990 NASA study showed that in 75 percent of the 3270 general aviation accidents that happened between 1983 and 1987, the ELT didn't even activate.

The National Transportation Safety Board, in a recommendation to the FAA, suggested that current ELTs are lacking in that they don't transmit on the international search-and-rescue satellite frequency of 406 MHz. The international satellite system can detect ELTs broad-

The ELT lies dormant within the plane waiting for that one moment when it is triggered by an impact with the earth.

casting on 121.5 or 243.0 MHz; however, it requires an additional pass by the satellite, delaying possible rescue. There is a TSO for 406-MHz ELTs (TSO C126), but there hasn't been a rush to build these units, probably because they cost much more to build, although Artex does offer one.

NTSB also recommended that FAR 43, Appendix D be amended to require mandatory periodic inspections of ELTs. This would include physical inspection of mounting, batteries for corrosion, antenna, wiring, G switch, and testing of the ELT's frequency alignment. Currently, the only requirement is that the ELT be working and that batteries be changed periodically.

TSO C91A applies to current ELTs and has standards for better G switches, antenna tethers, and stronger housings, which should reduce the amount of false alarms. All new ELTs must meet TSO C91A.

Artex now manufactures the ELS-10 ELT system, which it purchased from Arnav Systems. The ELS-10 interfaces with many LORAN and GPS systems and not only transmits the standard ELT signal, but also the location of the downed aircraft, based on LORAN or GPS information.

ELT Regulations

Title 14, Section 91.207 gives pilots some relief for an ELT that needs to be removed for repairs. Most people misinterpret 91.207 to read that they can only fly their airplane within a 50-mile radius after the ELT is removed. That, however, isn't the case.

First, 91.207 permits ferrying of newly acquired airplanes without ELTs to a place where the ELT is to be installed or ferrying an airplane with an inoperative ELT to a repair shop where it can be fixed. Here's the 50-mile rule: An airplane used for flight-training operations conducted entirely within a 50-nautical-mile radius of the airport where the training operations take place need not have an ELT installed at all.

Now here's the rule for ELTs removed for repair: FAR 91.207 (e)(10) states that for up to ninety days, an aircraft can be operated without its ELT, provided the aircraft records (logbook) contain an entry that includes the date of initial removal, make, model, and serial number of the ELT, and reason for removal. In addition, a placard must be installed within view of the pilot, stating "ELT not installed."

Thus, you can still operate your airplane normally while the ELT is being sent out for repairs. Don't forget that this removal can only be done by a licensed mechanic or under supervision of a licensed mechanic. If you are flying and you go down, it is highly unlikely you will be found alive without the Emergency Locator Transmitter.

FAR 91.207 Emergency Locator Transmitters

(a) Except as provided in paragraphs (e) and (f) of this section, no person may operate a U.S.-registered civil airplane unless—
- (1) There is attached to the airplane an approved automatic type emergency locator transmitter that is in operable condition for the following operations, except that after June 21, 1995, an emergency locator transmitter that meets the requirements of TSO-C91 may not be used for new installations:
  - (i) Those operations governed by the supplemental air carrier and commercial operator rules of parts 121 and 125;
  - (ii) Charter flights governed by the domestic and flag air carrier rules of part 121 of this chapter; and
  - (iii) Operations governed by part 135 of this chapter; or
- (2) For operations other than those specified in paragraph (a)(1) of this section, there must be attached to the airplane an approved personal type or an approved automatic type emergency locator transmitter that is in operable condition, except that after June 21, 1995, an emergency locator transmitter that meets the requirements of TSO-C91 may not be used for new installations.

(b) Each emergency locator transmitter required by paragraph (a) of this section must be attached to the airplane in such a manner that the probability of damage to the transmitter in the event of crash impact is minimized. Fixed and deployable automatic type transmitters must be attached to the airplane as far aft as practicable.

(c) Batteries used in the emergency locator transmitters required by paragraphs (a) and (b) of this section must be replaced (or recharged, if the batteries are rechargeable)—

 (1) When the transmitter has been in use for more than 1 cumulative hour; or

 (2) When 50 percent of their useful life (or, for rechargeable batteries, 50 percent of their useful life of charge) has expired, as established by the transmitter manufacturer under its approval.

 The new expiration date for replacing (or recharging) the battery must be legibly marked on the outside of the transmitter and entered in the aircraft maintenance record. Paragraph (c) (2) of this section does not apply to batteries (such as water-activated batteries) that are essentially unaffected during probable storage intervals.

(d) Each emergency locator transmitter required by paragraph (a) of this section must be inspected within 12 calendar months after the last inspection for—

 (1) Proper installation;
 (2) Battery corrosion;
 (3) Operation of the controls and crash sensor; and
 (4) The presence of a sufficient signal radiated from its antenna.

(e) Notwithstanding paragraph (a) of this section, a person may—

 (1) Ferry a newly acquired airplane from the place where possession of it was taken to a place where the emergency locator transmitter is to be installed; and

 (2) Ferry an airplane with an inoperative emergency locator transmitter from a place where repairs or replacements cannot be made to a place where they can be made.

 No person other than required crewmembers may be carried aboard an airplane being ferried under paragraph (e) of this section.

(f) Paragraph (a) of this section does not apply to—
  (1) Turbojet-powered aircraft;
  (2) Aircraft while engaged in scheduled flights by scheduled air carriers;
  (3) Aircraft while engaged in training operations conducted entirely within a 50-nautical mile radius of the airport from which such local flight operations began;
  (4) Aircraft while engaged in flight operations incident to design and testing;
  (5) New aircraft while engaged in flight operations incident to their manufacture, preparation, and delivery;
  (6) Aircraft while engaged in flight operations incident to the aerial application of chemicals and other substances for agricultural purposes;
  (7) Aircraft certificated by the Administrator for research and development purposes;
  (8) Aircraft while used for showing compliance with regulations, crew training, exhibition, air racing, or market surveys;
  (9) Aircraft equipped to carry not more than one person; and
  (10) An aircraft during any period for which the transmitter has been temporarily removed for inspection, repair, modification, or replacement, subject to the following:
    (i) No person may operate the aircraft unless the aircraft records contain an entry which includes the date of initial removal, the make, model, serial number, and reason for removing the transmitter, and a placard located in view of the pilot to show "ELT not installed."
    (ii) No person may operate the aircraft more than 90 days after the is initially removed from the aircraft.

The following is extracted directly from FAA's AC 43.13-1B, dated 9/8/98. This outlines the procedure for having the ELT (Emergency Locator Transmitter) tested.

12-21. EMERGENCY LOCATOR TRANSMITTERS (ELT). The ELT must be evaluated in accordance with TSO-C91a, TSO-C126 for 406 MHz ELT's, or later TSO's issued for ELT's. ELT installations must be examined for potential operational problems at least once a year (section 91.207(d)). There have been numerous instances of interaction between ELT and other VHF installations. Antenna location should be as far as possible from other antennas to prevent efficiency losses. Check ELT antenna installations in close proximity to other VHF antennas for suspected interference. Antenna patterns of previously installed VHF antennas could be measured after an ELT installation. Testing of an ELT must be performed within the first 5 minutes of an hour, and only three pulses of the transmitter should be activated. For example, a test could be conducted between 1:00 p.m. and 1:05 p.m., with a maximum of three beeps being heard on a frequency of 121.5 MHz.

12-22. INSPECTION OF ELT. An inspection of the following must be accomplished by a properly certified person or repair station within 12-calendar months after the last inspection:

(a) Proper Installation.
  (1) Remove all interconnections to the ELT unit and ELT antenna. Visually inspect and confirm proper seating of all connector pins. Special attention should be given to coaxial center conductor pins which are prone to retracting into the connector housing.
  (2) Remove the ELT from the mount and inspect the mounting hardware for proper installation and security.
  (3) Reinstall the ELT into its mount and verify the proper direction for crash activation. Reconnect all cables. They should have some slack at each end and should be properly secured to the airplane structure for support and protection.

(b) Battery Corrosion. Gain access to the ELT battery and inspect. No corrosion should be detectable. Verify the ELT battery is approved and check its expiration date.

(c) Operation of the Controls and Crash Sensor. Activate the ELT using an applied force. Consult the ELT manufacturer's instructions before activation. The direction for mounting and force activation is indicated on the ELT. A TSO-C91 ELT can be activated by using a quick rap with the palm. A TSO-C91a ELT can be activated by using a rapid forward (throwing) motion coupled by a rapid reversing action. Verify that the

ELT can be activated using a watt meter, the airplane's VHF radio communications receiver tuned to 121.5 MHz, or other means (see NOTE 1). Insure that the "G" switch has been reset if applicable.

(d) For a Sufficient Signal Radiated From its Antenna. Activate the ELT using the ON or ELT TEST switch. A low-quality AM broadcast radio receiver should be used to determine if energy is being transmitted from the antenna. When the antenna of the AM broadcast radio receiver (tuning dial on any setting) is held about 6 inches from the activated ELT antenna, the ELT aural tone will be heard (see NOTE 2 and 3).

(e) Verify That All Switches are Properly Labeled and Positioned.

(f) Record the Inspection. Record the inspection in the aircraft maintenance records according to 14 CFR part 43, section 43.9.

We suggest the following:

I inspected the (Make/Model) ELT system in this aircraft according to applicable Aircraft and ELT manufacturer's instructions and applicable FAA guidance and found that it meets the requirements of section 91.207(d).

Signed: ____________________

Certificate No. ____________________

Date: ____________________

NOTE 1: This is not a measured check; it only indicates that the G-switch is working.

NOTE 2: This is not a measured check; but it does provide confidence that the antenna is radiating with sufficient power to aid search and rescue. The signal may be weak even if it is picked up by an aircraft VHF receiver located at a considerable distance from the radiating ELT. Therefore, this check does not check the integrity of the ELT system or provide the same level of confidence as does the AM radio check.

NOTE 3: Because the ELT radiates on the emergency frequency, the Federal Communications Commission allows these tests only to be conducted within the first five minutes after any hour and is limited in three sweeps of the transmitter audio modulation.

Installation Tips

The criteria for ELT installations is somewhat specific. The ELT must be located somewhere in the aft fuselage where it can be securely mounted and the likelihood of the ELT surviving a crash is greatest. If existing shelving isn't available for mounting the ELT, the bracket provided with it can usually be mounted directly to the fuselage structure. The Narco ELT-10, used on Pipers for many years, is riveted directly to the aft fuselage skin, and that has proven to be a successful location. Wherever the ELT is mounted, the installation must allow the ELT to face in the proper direction, as marked on the ELT by a big arrow. Usually the ELT faces forward, but some are vertically mounted ELTs as well. This is important for proper operation of the ELT's G-switch. If this is a new installation, consider installing it closer to the cockpit, perhaps behind the baggage compartment, so you can reach the ELT from inside the airplane and get to it quicker in case of an accident.

However the rack is installed, make sure it is properly bonded to the airframe. This means you need good metal-to-metal contact between the skin and the bracket. Also recommended is corrosion treatment of the bare metal, preferably with Alodine 1200™ solution, before bracket installation.

The ELT's on/off/arm switch must be accessible to the pilot through an access panel that can be easily removed. In most cases, screws secure the access panel, so you should carry a straight-slot and Phillips-head screwdriver with you in the airplane at all times. In addition, new installations are now required to have a remote ELT switch in the cockpit. Most ELTs are already wired for this; it's just a matter of hooking it up and installing the wiring and the switch in the cockpit (see 91.207).

Because the ELT depends on the external antenna for efficient transmission of emergency signals, make sure the antenna is bonded properly to the airframe. Anything getting in the way of good metal-to-metal contact between the base of the antenna and the airframe, like rubber gaskets, corrosion, or paint can cause inefficient antenna operation. ELT antenna bases are usually made of stainless steel, so it's doubly important that the area where it will be mounted be treated with Alodine 1200™ to prevent dissimilar metal corrosion between the steel antenna and aluminum skin. Watch this area during your preflights, because it will corrode eventually, and clean up the corrosion before it becomes bad enough to hamper operation of the antenna.

To bond the antenna to the airframe, remove paint and any corrosion in an area 1/16-inch larger than the antenna footprint. Clean the antenna base and corrosion proof the area around the hole with Alodine 1200™ treatment. Install the antenna and check for resistance between the antenna base and the airframe of 0.0025 ohms or less. *Note:* you might need someone else to tighten the nut inside the airframe while the antenna is held, or vice versa.

After the antenna is installed, apply a bead of silicone sealant around the antenna base. Hold the sealer nozzle at a 70-degree angle to the skin and move the tip backwards to help prevent air bubbles from forming under the sealant. Then smooth the sealant down using your fingers or a small spatula to obtain a consistent 45-degree slope from the antenna base to the skin. The sealant should cover the area of paint you cleaned.

If the antenna is installed on a contoured area, you might have to use a contoured interface aluminum spacer that matches the footprint of the antenna. Failing that (high cost), you'll have to add Alclad aluminum, bonding washers to make contact with the skin. Another alternative is an Alclad gasket, made of soft aluminum that will forms or crushes to the contour when you tighten the antenna down. Don't discard any gaskets that the vendor requires be installed with the antenna. Again, I recommend the application provided by Dayton Granger that uses a process called Ultra Seal. This solves all your problems and then some. The cost for the sealer is about $45, somewhat expensive for most of us. There is an alternative, but you will have to contact the author of this book for more information.

With some antennas, the mounting bases are nonconductive, and vendor-supplied conductive foam gaskets are required.

ELT antennas are best located at least 3 feet from other antennas. On Piper Navajos and some other aircraft, the ELT is located inside the dorsal fin, which allows RF to pass through without hindrance and provides good protection from weather conditions that could harm the ELT. Wherever it is located, it should never be exposed to the elements.

Because ELTs operate on so few frequencies, only an inexpensive whip antenna is needed, usually provided with the ELT. If your antenna breaks, which isn't unusual, make sure you buy the antenna specified for the aircraft's ELT, not just any old antenna that fits the existing hole.

Battery Replacement

Every time I've shown owners what their ELT battery looks like, they always ask why they cost so much. The battery packs generally look like a bunch of D-cell alkaline batteries wired together. Essentially, that is what they are, except they are wired internally in a special way that is approved by the FAA and that allows the batteries to have a 2- to 3-year operational life in your ELT. Normal batteries (cells) have a pressure contact internally, whereas the ones required by the FAA are soldered to the cap instead of relying on a mechanical contact.

The only way to assure high reliability of ELT batteries is for the battery manufacturer to procure raw batteries and weld all connections to form a dependable battery pack. All cells are mated into one unit using nickel-plated straps, and the entire assembly is inserted into a plastic case or mold-epoxied to prevent relative movement between individual cells. Teflon wires are soldered to the positive and negative terminals of the battery pack, and high quality AMP or Molex® mating connectors are attached for connecting the battery to the ELT.

Shelf lives for batteries vary, depending on type of battery used, design of the battery pack, and quality control during assembly. The alkaline cells used by ELT battery manufacturers have about a six-year shelf life, but for replacement purposes, the FAA cuts that down by half, to a maximum of three years. You'll find you can buy either two- or three-year battery packs for your ELT.

When buying a new ELT battery, check the replacement date marked on the battery. If you bought a two-year battery, your new battery should have a solid two years left. Don't accept a battery whose life is already short by a few months.

To change the ELT battery, you must first remove the ELT from the airplane. Remember that this is something you can do only under the supervision of a licensed mechanic. Open the ELT access panel, making sure at the same time that the panel is easy to open. You don't want to find out that the corroded screws on the access panel won't come off the day you have a survivable accident.

Before attempting to remove the ELT, turn the on/off/arm switch to the "off" position. You could easily activate the ELT by removing it if you left the switch in the "arm" position. The ELT is usually held on its bracket by a simple clip; this is used on Narco's popular ELT-10, found on most Pipers. Many Cessna aircraft have the Dome and Margolin ELT installed, with four Phillips-head screws holding it on. Cessna usually mounts its ELTs on the upper fuselage just aft of the baggage compartment.

To complete the removal of the ELT, you'll have to unscrew or disconnect any wires that go to the remote switch and the antenna coaxial cable. Don't mix up the remote wires, and mark them if you're not sure you'll be able to tell where they go back on. Note that on the ELT-10, there is a little plastic tab that sits between the ELT's own portable antenna (die flat, flexible metal strap wrapped around the antenna) and the ELT's plastic body. This tab, usually attached to the main antenna coax, must be installed between the portable antenna and the ELT whenever the main antenna is attached. That way, the portable antenna is disabled when the main antenna is attached, but if the unit is removed, the portable antenna is automatically connected. Then, all you have to do is extend the antenna strap and you can transmit as well as you could with the main antenna in the airplane. This is useful if you have to leave the airplane and want to take the ELT with you.

ELT Testing

According to the FAA *Airman's Information Manual,* testing of ELTs, if necessary, should be done at five minutes past the hour and for a maximum of three audible sweeps. Testing at the wrong time and for too many sweeps could lead to unnecessary search-and-rescue efforts, so keep testing to a minimum. The reason for the specific testing time is the FAA will be expecting those tests at that time, unless they continue without stopping. Then they would be considered a valid emergency.

There are two tests that you can perform. The first is to tune your radio to 121.5 MHz, then flip the ELT switch to the "on" position, then "off" after you've heard three or fewer sweeps.

The second test is to make sure the G-switch is working and the ELT switch is in the "armed" position. Again, have your radio tuned to 121.5 MHz. Holding the ELT firmly, quickly move the ELT in a horizontal plane in the direction of the arrow printed on the ELT's body. Stop the movement suddenly in order to impose a G-load on the ELT. This should set off the ELT. Shut the ELT off before more than three sweeps are heard. Don't try to test the G-switch by whacking the ELT—it's not good for the unit's electronics.

Do not forget to leave the ELT with its switch in the "armed" position after reinstalling it. Otherwise, it won't go off in a crash.

ELT Repairs

Usually the best company to fix your ELT is the original manufacturer. Manufacturers offer reasonable turnaround time and also check the ELT for any other problems or updates that might be necessary.

ELT Manufacturers

These ELT manufacturers' addresses are listed in the Sources section of this book.

ACK Technology	Emergency Beacon
Artex Aircraft Supplies	Narco Avionics
Dome and Margolin	Pointer

CHAPTER 18

Radar Systems

More and more smaller airplanes are showing up with radar already installed, and some even feature both a radar- and electrical-discharge-detecting Stormscope®. Radar is still much more expensive than the Stormscope® and costs more to maintain, because of moving parts like antenna bearings that wear out, and other expensive components like waveguides and magnetrons.

All radars have included the following basic components:

1. **Indicator.** This is what displays the weather information. It is basically a CRT screen enclosed in a housing and sized to fit your particular installation.
2. **Receiver-transmitter (RT).** This includes the magnetron that generates the X-band radar signal and the circuitry that evaluates the returned signals and drives the indicator display.
3. **Waveguide.** The radar signal must travel through the waveguide to reach the antenna. The waveguide is like a high-tech pipe through which the X-band signal flows, and it is necessary because regular wire cannot carry the signal.

4. **Antenna.** The antenna sends the radar signal to the target, usually raindrops in a cloud, and collects the signals as they return from the target for processing by the RT.
5. **Radome.** This provides an aerodynamic cover for the antenna that is transparent, or nearly so, to radar signals.

Radar Antennas

The radar antenna, whether mounted in the nose of a light twin, in a pod under the wing of a single-engine airplane, or even in the leading edge of the wing, is a complex piece of equipment. When equipped with gyro stabilization, the antenna rotates up and down and side to side to compensate for the pitching and rolling of the aircraft, thus keeping the same area in radar view, within limits, as the airplane turns and climbs. Without gyro stabilization, the radar antenna still moves as commanded by the pilot, and during all this movement, RF energy travels from the RT to the antenna through the waveguide then out to the target. After arriving at the moisture-laden clouds, the RF energy is reflected back to the antenna, travels through the waveguide and into the RT, where the radar display shows the pilot what lies ahead.

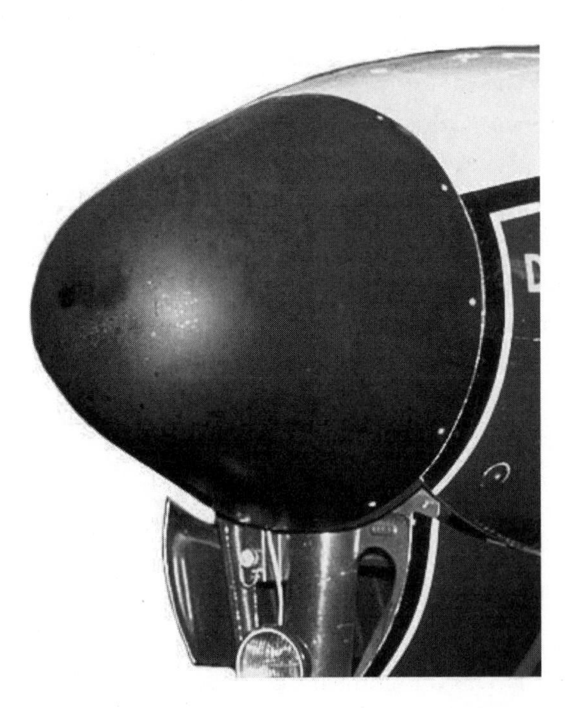

Radar radome.

All of this action depends on the correct action of resolvers, motors, relays, and amplifiers. Should any contaminants such as moisture, dirt, or other corrosive contaminates enter the RT or the driving gear and motors, the radar's operation will become erratic or simply fail altogether. Proper preventive maintenance will prevent contaminants from affecting radar operation.

Radar reflector antenna.

Most radar antennas used on general aviation aircraft today are the flat-plate, phased array, radiator type. The power efficiency of the flat-plate antenna is significantly greater than the parabolic antenna because of lower side-lobe loss. Power from the RT is distributed to the right and left sections of the antenna, to be dispersed through slots in the antenna. The slots are slanted at varying angles from the centerline of the waveguide. The RF energy is "guided" from the waveguide through the rear of the antenna. As the energy passes through the slots in the antenna, there is a phase shift, but each slot has an opposing slot that acts to rephase the energy. The RF emerging from the front of the antenna is in phase, and accordingly, so will be the reflected signal.

Parabolic antennas are still used in larger commercial aircraft because for the size of the antenna, the weight is less than it would be for a comparable-size flat-plate antenna. Less tilt is needed for ground mapping with a parabolic antenna, plus the larger the parabolic dish, the greater the efficiency advantage over flat-plate antennas. A flatplate antenna will still operate with dents or dings around the edges, but if a parabolic antenna is dented, the radiated efficiency can be drastically reduced.

Radome

The radome is a streamlined, RF-invisible cover designed to protect the radar antenna while presenting as streamlined a profile to airflow as possible. Because the radome is made of materials like fiberglas, which radio waves can pass through without hindrance, radomes are more sensitive to abrasion and cracking than a comparable metal surface. It is important to inspect radomes regularly for cracking and delamination to prevent premature failure, which could allow moisture to enter the radar antenna and damage moving parts or the radar itself.

Radomes are tuned to radar frequencies and "tweaked" for beam-to-wall incidence angle. The tweaking is essentially designing the walls of the radome to handle structural loads but still allow radar beams to escape and enter with minimum loss of efficiency. Although the thickness of the radome might not seem excessive, remember that the radar beam exits at an angle. This angle could cause the beam to have to travel through a thickness four times that of the actual thickness of the radome wall.

Not all radomes are round cones found on aircraft noses. You'll see small radomes on wing-mounted radar pods on high-wing single-engine aircraft like Cessna's 210 and turbine-powered Caravan. On Piper's single-engine piston-powered Malibu, the radome is actually part of the leading edge of the wing, and it is protected by a portion of the deicer boot. Interestingly, Goodyear, the manufacturer of the deicer boot, makes a boot out of a special type of rubber for the right wing of the Malibu that is more suited to efficient radar transmission. This boot is different enough from the "normal" type of boot used on the left wing that after a few years of operation, you can spot the difference in wear and tear between the left- and right-wing deicer boots on a Malibu. The left boot, after about five years, will have lots of cuts, pinholes, and patches, while the right boot will look relatively clean.

RADOME PROBLEMS

Improperly designed or repaired radomes can set up superfluous reflections, causing undesirable side lobes of radar energy to radiate from the antenna. Of course, these undesirable reflections always seem to occur at the worst possible time, causing random and less than accurate returns. If they happen during adverse weather conditions, they can obviously increase pilot workload. Because reductions in radar efficiency could be a result of radome problems, technicians might be unable to duplicate the problem. This might leave you with ineffective radar just when you're trying to avoid a thunderstorm. Don't ignore the radome as a source of possible radar trouble.

RADOME ABUSE

Things people do to their radomes: Painting with the wrong type of paint can destroy a radome's electromagnetic transparency, plastic shields can cause beam distortion and create back-reflections that generate ghost storms. These same shields can also make it more difficult for the radar to discriminate intensity levels and prevent the radar system from contouring some storm cells.

A good radome will lose about 20 percent of the reflected and transmitted power. Some abused or poorly maintained radomes can have as much as a 50 percent power loss.

There can be no holes or cracks on a radome, otherwise moisture can enter and slowly move from cell to cell, eventually causing very expensive damage. This can be accelerated by freezing. The water freezes and cracks an adjoining cell, water enters, and the process continues, creating a domino-like effect until many cells are ruptured.

If the pilot should see a hole or crack, he or she should immediately cover it with duct tape and get to a shop for repair. Always inspect the radome for smudges, scrapes, small burn marks, and holes. Quick response may save several hundred dollars.

Anyone can check the radome for delamination or voids by using a quarter and tapping on the surface of the radome while listening for either a resonant, lively response or a dull, non-responsive reply. The dull reply indicates of delaminated core from the skin.

Do not paint the metal static/lightning discharge strips that run along the surface of the radome. When struck by lightning, the painted strips will evaporate and the radome is history. Unpainted strips will shunt the strike off to the more protected fuselage skin, as designed.

You can dynamically check your radar system by performing a simple ground-map test. At a distance of 100 miles, the radar screen, when you have the antenna tilted down to map the ground, should show normal display of ground features to the left, right, and directly ahead of the airplane's path. Irregular display scans or dead portions of the scan might indicate a problem with the antenna system or radome. What is important to remember is that the radome is as much a part of the radar system as the RT or CRT. If the radar system can't be easily pinpointed as the cause of the problem, then the trouble is most likely in the radome.

The easiest way to rule out a suspected radome is to ship it to an authorized radome repair facility, especially for a radome that has been in use for a long time without repair. In fact, the repair facility might find other problems with the radome that you weren't even aware of, so be prepared for that possibility. If you want a free check of your radome, ask your dealer for the location of the Akron, Ohio, Norton radome facility. If they are not overloaded, they will gladly check your radome.

Too often, the radome is taken for granted. Everything from small stones to birds has its way with the radome. Receiving a torrent of aerial contamination, the composite radome is abused repeatedly. Many times this "normal" wear and tear creates a potential failure point, through which moisture can enter and freeze at altitude. As this takes place, the bonded layers of fiberglas weaken until they delaminate.

During scheduled inspections such as annuals, these bruised areas are often overlooked or ignored because of cost considerations. Although the customer should be apprised of the damage, there are many times they aren't and the damage remains. Or when the owner is told about the problem, repairs might be put off because of high costs. Contributing to the problem is that ground personnel frequently push on radomes when moving airplanes. And finally, add to that wear and tear from radome removal for radar and other nose-mounted avionics maintenance.

During your preflight inspections, make it a habit to look closely at the radome. The earlier you find a damaged radome, the sooner you'll be able to get it fixed and keep the radar in top operating condition. Be especially alert for unreported damage that has been repaired incorrectly. Radome repair is an exacting and detailed field, and not just any mechanic can repair radomes properly. Unfortunately, there have been cases of radomes that were damaged during ground handling or were repaired incorrectly, causing poor radar operation. Some of the evidence to look for of improper repairs includes discoloration, deformation, large concave areas, and cracking.

Here is a list of what mechanics should look for when inspecting a radome during a routine annual inspection:

1. Scuffing, chipping, peeling, or painted surface.
2. Surface cracks (microcracking in composite material).
3. Low areas, especially where surface cracks are evident in circular patterns.
4. Damaged mounting holes (fiberglas torn).
5. Water oozing from cracks (water has entered fiber core).
6. Lightning-strike diverter damage.
7. Bonding strips damaged.

8. Excessive erosion wear.
9. Small pinhole-sized burn marks.
10. Poor paint job, signs of cratering, or pinholes.
11. Color change of erosion/static coatings from black to brown.
12. Plastic erosion cap improperly installed.

If damage has occurred, look for signs of water contamination. This might look like a white, crusty trace or black streamers radiating away from a source of air flow. The white, crusty trace is mineral deposits left after water evaporates, and these deposits are especially bad for antenna bearings. The black streamers are a combination of oil and dirt, also not a good lubricant for moving parts. Failure to inspect regularly for this type of damage could result in failure caused by corrosion.

Radar efficiency depends on a tight, narrow pulse of intense electromagnetic energy leaving the antenna, arriving at a target, and being reflected back to the antenna. If the energy is prematurely reflected, diffused, and spread out as it leaves the radome, a weaker return is the result, and the pilot gets erroneous information on the radar screen. Should water get trapped between laminated areas of the radome, it will effectively block the radiated signal as it leaves the antenna and attenuate the signal when it returns from the target.

Detection of trapped water is difficult, especially when it comes and goes as the temperature of the radome changes with attitude. Few shops have the equipment on hand to check for moisture contamination, although certified radome repair shops check for moisture as a normal part of a repair.

One mistake to watch for occurs during repainting. Some paint shops apply striping along the side of the fuselage, terminating in a neat, narrow taper at the front of the nose, on the radome. Besides presenting a thicker path to the radar signal, these stripes frequently contain metallic particles, for that distinctive "look," which can easily reflect or diffuse radar signals. The distortion caused by the degraded signals could be extreme enough to show up on the radar screen.

Another painting mistake is failing to remove the original paint and just painting over the existing radome. The added paint layer can radically decrease signal efficiency. Even removing the old paint for repainting is a delicate operation. One wrong swipe with a sanding block, and too much of the radome material can be removed. This material must be replaced to retain the radome's design configuration and aerodynamic shape.

Your best bet when having your radar-equipped airplane repainted is to find out if the shop knows exactly how to handle radomes. If not, have your radome sent out for overhaul while your airplane is being painted, and tell the radome shop what colors are being used on your airframe so it can match the colors if this can be done without degrading the radome's performance.

Scuff marks on the radome usually indicate items striking the radome in the air or while taxiing on poor surfaces, or from hangar rash. "Hangar rash" is a nickname for mishandling of the airplane on the ground by maintenance handlers that results in dings and nicks and scrapes from running into hangar walls, other aircraft, or equipment. Don't be too worried about scuff marks on the radome unless other signs of damage are also visible. If scuff marks are accompanied by circular cracking, try pushing gently on the center of the circular area. If the section pops inward, there's a good chance the outer surface of the radome is damaged, and it needs to be inspected closer.

Paint chipping is usually the result of impact from ice or birds, from heat buildup from static charges, or poor paint application. When you spot paint damage such as chipping or peeling, have it repaired quickly. Delaying such repairs could allow moisture entry and freezing in the fiberglas radome. Like the small nick in an auto windshield, paint damage can develop into far greater damage than originally suspected. Small pits or nicks from sand or other fine particles can be filled with a clear acrylic paint until more complete repairs can be made.

When humidity is low there is a greater chance of static electricity discharge occurring. As air moves across the surface of the radome, electrons are stripped off the air molecules, leaving a negative charge. Often this charge is quite high and can cause radio problems or even damage to the radome's paint as the charge attempts to leave the radome and migrate to the airframe or into the atmosphere. This problem is most prevalent on high-speed aircraft such as heavy turboprops and jets.

To minimize the static problem, antistatic/antierosion coatings are applied to the nose of radomes. Although these coatings reduce the effects of static, they don't eliminate the problem. The black coating should last at least two years; after that the coating will turn gradually from black to brown, at which time it should be replaced by an authorized facility. Recommended coatings are either a black neoprene or polyurethane. Of the two, polyurethane is the best because it has very good antistatic and antierosion capabilities while retaining reasonable transmission efficiency.

Erosion caps are an alternative to paint-type coatings. These look like giant, soft contact lenses and are made of a thin film of formed polyurethane with an adhesive backing. Erosion caps have virtually no resistance to radar beams. The thicker, solid plastic caps aren't a good idea because it is difficult to obtain positive cap-to-radome contact, and this could allow moisture to get trapped behind the cap, degrading the radar signals.

Lightning-strike diverter strips are an important radome component and should be inspected regularly. The heat generated within the nose cone by a lightning strike can be enough to blow the nose apart. Diverter strips carry charges away from the nose to the airframe. Although lightning can be destructive, much of the damage can be reduced if diverter strips are solidly bonded to the airframe. Static wicks are available that can handle both static discharge and lightning.

RADOME REPAIRS

The simplest type of radome repair is injecting resin into pinholes caused by static burns. Technicians use a hypodermic needle to inject epoxy polyester into the hole and to fill open areas around the hole. Prior to the injection of epoxy, any carbon residue from the burn should be removed because leftover residue could act as a conduit for electrical charges, causing more damage. Adding too much resin can affect radar performance, so with this and any other radome repair, you should take it to a radome expert.

While composite repair principles apply to radomes, at least insofar as they are made primarily of composite materials, the tooling, experience, and skills to return the radome to original design configuration aren't found in most FBOs. Specialized radome repair shops have equipment for detecting moisture contamination, and after repairing a radome, these shops test the radomes to ensure they are within design limits for efficient radar transmission. All this equipment and expertise doesn't come cheap, but it's worth it to make sure your radome is providing the best protection for your radar antenna and optimum transmissibility for the radar beams.

Here is a case where lack of attention to proper radome repair might have been a factor in a fatal accident. It involved a twin engine plane on an instrument flight whose pilot was relying on radar to avoid severe weather. In reality, a severe weather front with huge thunderstorms lay directly in the pilot's flight path, but it seems that the pilot didn't spot the severe weather on the radar. The pilot was well-qualified in the aircraft and its radar system and knew better than to try to fly through a storm of that magnitude. It can only be speculated that the pilot was totally unaware of the storm's intensity and proximity and was well into it before being able to turn away. But how did this happen to a conscientious pilot?

The answer could lie in what occurred earlier. While the airplane was being towed out of its hangar, it struck a tug, severely damaging the front of the thin-walled radome. A field repair was performed on the radome, and the airplane was returned to service. What the pilot didn't know was that the radar system would display returns from the left and right of the airplane fine, but would indicate less weather than was actually directly in front of the aircraft. Two flights later, the airplane came raining out of a thunderstorm in pieces. Was the repair at fault? It's up to you to decide, but consider also the question, "Is an unauthorized repair worth the risk?"

The world of radomes is quite complicated. Even the original specifications issued by the Radio Technical Commission for Aeronautics are being modified because of differing opinions about transmission standards. A new spec for radomes is being written, with two separate classes:

1. Type A, predictive windshear.
2. Type B, covering conventional weather radar.

Because radar manufacturers are motivated to provide the customer with the most power possible with easy-to-interpret displays, they are requiring that radomes meet the highest possible specifications. Honeywell, for example, requires a maximum 10-percent transmission loss for radomes to be used with its radars.

Installation Tips

Vibration is an enemy of radar. The radar must not be mounted too close to the aircraft's skin; otherwise, the buffeting of the skin against the radar RT or the antenna dish may cause damage. High-frequency vibrations can be created by odd combinations of bracketing and shelf attachment but can be avoided if the radar is properly mounted using vendor recommended installation procedures and the correct hardware.

Corrosion and Moisture Contamination

Radar's worst adversary is moisture, which causes corrosion. Corrosion is normally the result of poor design, installation, or maintenance and can cause expensive problems, sometimes in the range of thousands of dollars. Have your maintenance facility properly seal the radome and place a moisture detector in the compartment. Periodically, especially after a heavy rain, check it for signs of moisture intrusion.

The RT and its magnetron, the power supply, circuit boards, antenna bearings, and attaching hardware are all subject to failure from corrosion. Corrosion causes increased resistance in electrical connectors and printed circuit board contacts, drastically altering the operating parameters and overloading critical circuits. In addition to electronic failures, water and other contaminants can enter the antenna drive bearings, reducing their efficiency and causing eventual failure if preventive action isn't taken.

The evidence you're looking for is a black, graphite-colored substance and a white mineral deposit, which is a combination of oil, moisture, and dirt particles. The white mineral deposit indicates that water has recently evaporated, leaving behind a crust-like substance that will act more like an abrasive than a lubricant, especially when it gets into antenna drive bearings. Moisture is a killer, especially when it acts as a transportation agent for hybrid contaminates such as oil, metal filings left over from a sloppy installation, dirt, and other corrosive elements. These can attack the circuitry within the radar or cause temperature shock as it comes in contact with electronics operating at high temperatures.

Waveguides must be kept clean and dry. Pressurized aircraft usually have provisions for pressurized air to be ported to the waveguide. This helps the waveguide operate more efficiently at high altitudes. But because pressurized air is frequently full of moisture, it is necessary to keep desiccant materials in the waveguide to absorb moisture. The desiccant should be replaced periodically.

Radar Squawks

Radar problems come in a confusing variety, and correct interpretation will help you narrow the problem down so you can save money on troubleshooting and get right to the source of the problem. This section describes some of the most common problems.

- CRT is not painting. The screen periodically shows only one sweep mark, it paints a scrambled screen, or it displays a target, then erases it. This could be a memory problem, or the radar might not be updating.
- If the target is displaced after each sweep, the radar could be out of sync, but it is still attempting to update. Multiple targets and/or sweep lines might also be displayed with this problem.

Radar system undergoing bench test. (Photo provided by Skywest Airlines.)

- If the scope isn't displaying weather, try initiating the test function by selecting test mode.
- If a test pattern is present, this indicates the data lines are intact, the system trigger is presently arriving at the RT, and the four-color levels are being correctly processed.
- If there is no test pattern, look for an error in the system trigger circuitry.
- If the CRT doesn't light up, expect the problem to be in the RT. If the CRT does light up and the range marks and alphanumeric symbols are being displayed but no test pattern is evident and the scope isn't painting weather, then look for system trigger failure.
- If the radar is transmitting and "spokes" are being displayed on the screen, the automatic frequency control (AFC) could be unlocked. The presence of the spokes indicates the transmitter is transmitting, but because the AFC is unlocked, only intermittent reception of the incoming RF is processed, only when that RF is within the frequency of the receiver. The fault most likely lies in the magnetron not being within its designated frequency range (most magnetrons generally must fall within a frequency range of ±35 MHz).
- When an antenna problem develops, it might show up as a jerky update sweep, smeared targets, or azimuths being clipped and

only painting one side. When the sweep reference is passing center (the bore sight), it updates itself. If the bore is out of adjustment, then "catch up" occurs, causing a smeared image. If the bore sight isn't calibrated correctly, it needs to be matched to the radar indicator's sweep line.

- Sophisticated radars have a gyro-stabilization system that keeps the radar horizontal even if the aircraft rolls. Proper orientation of the roll gyro for the radar's antenna stabilization system is critical for accurate radar scanning. If the radar is displaying strange patterns while rolling and pitching the aircraft, see if the patterns disappear when the airplane is held in a straight-and-level attitude. If so, this could indicate a failure of the stabilization system.

- If the radarscope displays excess ground clutter, either evenly centered or in the lower right or left corners of the screen, the antenna might require physical adjustment or there could be a gyro-stabilization error if a stabilization system is installed. The gyro may be mispositioned or could have failed electronically. The stabilization gyro must be aligned with the aircraft in the in-flight straight-and-level attitude and be positioned correctly fore and aft. Don't overlook the fact that the tilt adjustment could have been set too low, which would cause the radar to scan the terrain and cause the evenly centered ground clutter. This is an operational problem, not something wrong with the radar. For the problem where the radar paints one side or the other, you might suggest the technician perform a functional test on the gyro. If the gyro is replaced, expect the shop to perform a full radar stabilization alignment on your aircraft.

- If the CRT paints red all the time, look for a defect or misalignment of the tilt drive.

- Scrambled displays during communication transmissions aren't common. If you experience this problem, poor harness routing, open shielding and poor bonds could cause it. The first step is to try to isolate the problem by temporarily substituting COMM coax cables with known good test cables. Or try switching COMM 1 coax with COMM 2 coax. If the problem moves to the second COMM, the problem is in the communications antenna and/or coax cable. Bonding of the COMM antennas, instrument panel, the radar indicator, and installation of shielded wire are other steps that should help solve this kind of problem. The problem could be the radar indicator itself. It could be missing modifications that are designed to prevent the high-gain section of the radar from picking up extraneous RF from the COMM radios. If this is the case, it should be easier to simply have the radar indicator brought up to date instead of messing with coax and bonding and shielding, not that those factors shouldn't be done properly to start with.

- Don't forget the corrosion problems mentioned earlier. If you spot any corrosion, make sure it's taken care of right away before it causes radar problems.

Safety

There is much that pilots and technicians need to know about proper radar operation, and one of the best ways to learn about radar is one of Archie Trammel's radar seminars. For more information on seminar dates and costs, contact AGT, Inc. at RD 1 500 Rosemary Drive, Trinidad, TX 75163. The phone number is (903) 778-2177.

Radar Manufacturers

These radar manufacturers' addresses are in the Sources section at the end of this book.

Bendix/King

Collins Avionics

Honeywell

Narco Avionics

CHAPTER 19

Thunderstorm Detection Systems

Pilots have known for years that an ADF can show the location of thunderstorms. Also well known is that radar's accuracy is limited for thunderstorm detection. Radar displays any weather containing moisture, which may or may not be associated with thunderstorms. Fortunately, a device that can "see" the noise generated by lightning strikes using similar principles, as the ADF is available. This device makes it possible for a pilot to spot thunderstorms and avoid the hazards of convective windshear and turbulence without having to install a complex and expensive full-blown radar system.

Strike Finder® and Stormscope®

There are two such devices: One is the Strike Finder®, and the other the Stormscope®. Both have the same mission, to locate and display thunderstorms providing the pilot an opportunity to circumnavigate around it. As similar as their respective goals are, the path they take is as different as oil and water.

Strike Finder®

The Strike Finder® is manufactured by Insight Instrument. It is the first weather avoidance system to utilize digital signal processing technology. It models convective activity based upon the electromag-

Strike Finder® side view.

netic signature of lightning. Digital technology enables Strike Finder® to locate thunderstorms with greater accuracy and reliability than previously possible.

The Strike Finder® Digital Weather Avoidance System detects and analyzes the electrical activity emanating from thunderstorms within a 200-nautical-mile (nm) radius of the aircraft. A unique gas plasma graphic display plots an accurate, reliable and easily interpreted real-time picture of electrical activity that can be used to navigate around potentially hazardous thunderstorms.

Besides displaying location, the system also analyzes the individual strike signal properties to determine the bearing, range, and severity of the activity. Strike data is plotted on the display as single orange dots by range and azimuth, in relation to the aircraft symbol. As the number of lightning strikes increase, so does the number of plotted strike dots. Cells start to form indicating increased lightning activity.

Digital technology provides superior signal fidelity for a truer picture of weather activity, and it is free of false indications and gross errors in bearing and range caused by poor signal processing. This is unlike other storm avoidance systems, where errors in signal computation develop rendering the display virtually useless for accurate decision making. The Strike Finder® provides a level of security far beyond radar. Radar is no substitute for Digital Weather Avoidance since radar relies on the presence of rain for detection of thunderstorms. Strike Finder® routinely identifies hazardous convective activity long before any rain has fallen and before radar shows any returns.

A zoom feature, at the touch of a button, provides the pilot an accurate overview of weather activity within a 200-mile radius. Strike Finder® can display 200-, 100-, 50-, and 25-mile views for greater detail.

A unique feature is the ability to review past and present weather activity in time-lapse. Recent electrical activity can be replayed in a time-compressed format to reveal weather trends. This time-lapse picture of weather activity helps the pilot forecast weather conditions. These two features make the Strike Finder® the choice for many pilots because of a feeling of comfort and ease with the system and how it works.

Although the gas plasma display technology seems less state-of-the-art, it provides a highly readable graphic presentation of weather information at a lower cost and greater reliability. Gas plasma was chosen over CRT technology for its superior bright-light readability, size, simplicity, and long-term reliability.

Onboard computer diagnostics constantly monitor the operating performance of the complete system, providing weather avoidance technology the pilot can count on.

Direct slaving to King HSI compass systems without the expensive synchro option is one of Strike Finder's many cost savings. It also interfaces with any compass system providing standard synchro output.

Because there is no remote box, no troublesome interconnections, and a fraction of the components found in other passive weather avoidance systems, there is a greater degree of reliability and security that what you see is what you get.

Strike Finder® will never give the pilot deceptive displays because of a misalignment because there are no adjustments—not in the factory or the field! To the aircraft owner, this means you'll never have to tear your panel apart for periodic service.

Strike Finder® has its processor incorporated within the display case, which makes it simpler to install. Its range is similar to that of the WX-1000, but it is smaller and easy to install. The Strike Finder® is quite light, thanks to its orange plasma display flat screen and self-contained processor.

Out of the box, the system comes with a prebuilt harness on which only one end must be terminated. Because of the Strike Finder's compact configuration and unique processing design, it is described as being less susceptible to noise interference than the Stormscope®, which was designed to receive frequencies in the 50-kHz range (slightly below ADF generators). In fact, the Strike Finder® is virtually immune to onboard noise generators; it was designed to look at a broader frequency spectrum and detect only the sharp transients that form the activity inherent in lightning.

The Strike Finder® is so precise that the system can tell the difference between reflected radiation and that which is the direct result of a lightning strike.

INSTALLATION TIPS

For installation guidelines, refer to Stormscope's installation recommendations. The antennas are essentially installed the same, however, there is only one connector end to solder on the Strike Finder®.

Break out the champagne: Insight's revolutionary new stabilization module eliminates the need for a slaved compass system.

Until now, weather-mapping systems required they be connected to the aircraft's slaved compass system to maintain correct data orientation through heading changes. The problem is many planes don't have a slaved compass system and this new module resolves that obstacle.

The stabilization module provides data to the Strike Finder®, enabling displayed lightning strike information to rotate relative to

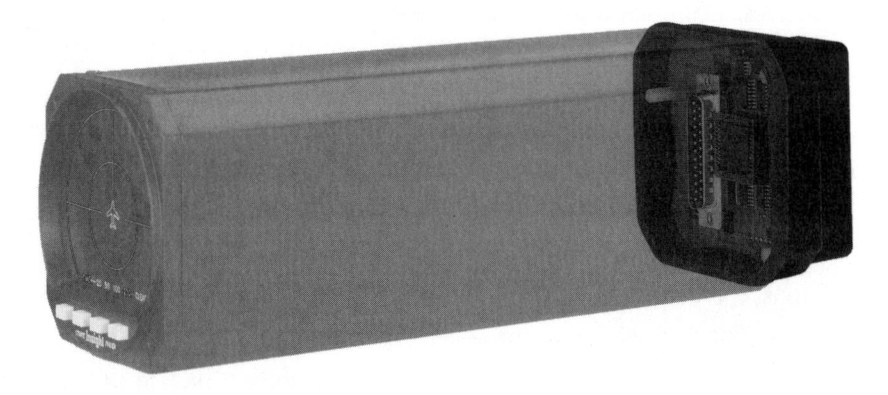

Strike Finder® stabilization module.

heading changes. Operation of the stabilization module is automatic; no field configuration or calibration is required. The revolutionary stabilization module

1. Eliminates the need for a slaved compass system.
2. Is totally self-contained sensor and data processor.
3. Simply plugs into the back of the Strike Finder® display.
4. Contains no rotating gyro.
5. No longer needs periodic overhauls.

LATEST TECHNOLOGY

The stabilization module contains all solid-state components; there are no moving parts. New software algorithms process data from the enclosed sensors to determine heading change. Insight has developed this state-of-the-art stabilization system by integrating both the motion sensor and the data processor into one miniature module. Benefits include reduction in cost, complexity, and weight compared to a conventional system.

FACTORY OPTION

As a factory-installed option, the stabilization module is installed within the Strike Finder® housing, deriving its power internally from the Strike Finder® and providing its signals directly to the Strike Finder® processor. Insight recommends the Strike Finder® system with the factory option, even for slave-compass equipped aircraft.

This installation is simpler, faster, and offers enhanced reliability by eliminating dependence on the maintenance-intensive rotating gyro.

UPGRADE OPTION

As a field upgrade, the stabilization module is provided in a housing, which snaps onto the rear of the Strike Finder® display. The Strike Finder® system upgrade option offers the exact same benefits as the factory option.

PLUG AND FLY!

All required power and signals are provided through the connector on the stabilization module for both internal and external operation. Installation is simple and fast.

FACTORY-CALIBRATED

Each stabilization module is individually factory-calibrated and tested to insure proper operation under all conditions.

Stormscope®

B.F. Goodrich's Stormscope® helps thousands of pilots safely avoid the hazards of convective weather. They spot dangerous conditions that radar might not show. The Stormscope® series enables pilots to make informed, flight-critical decisions about thunderstorm avoidance and fly with greater confidence. The Series II Stormscope® products incorporate the latest ranging technology for more precise tracking of lightning discharges from as far away as 200 nm.

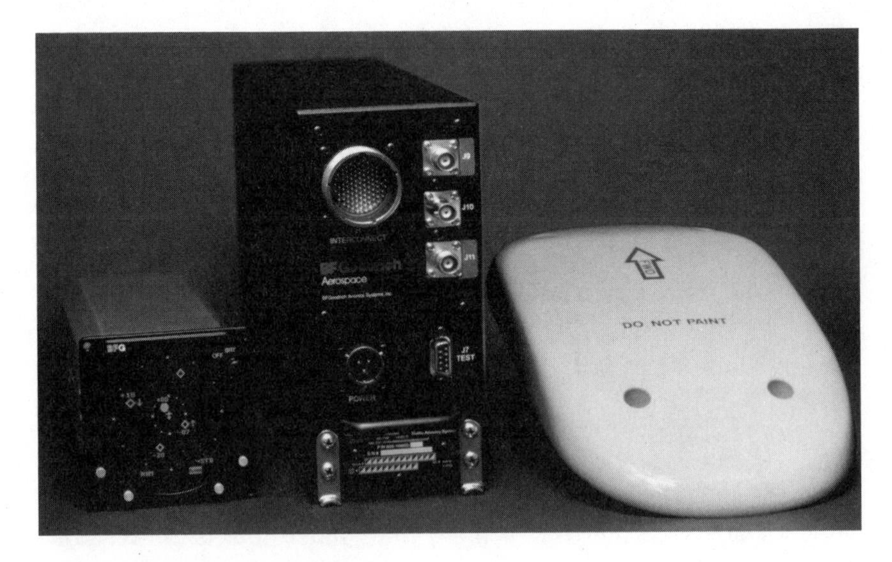

Stormscope system WX-500 receiver and antenna.

WX-500 SERIES

Tapping into the future, the WX-500 has a weather-mapping sensor that adds storm avoidance capability to your multifunction display.

WX-900 SERIES

This is a feature-rich, economical thunderstorm avoidance device featuring a liquid crystal display.

The lower-cost WX-900 lacks the power to detect lightning over 100 nm from the aircraft. The WX-1000 has a 200-nm range, which is practical during long-range flights where storms are in dynamic flux. If a storm were already brewing before a flight, the 100-nm range of the WX-900 would be satisfactory, especially with a slower-moving aircraft. The faster the airplane, the greater the need for a longer-range lightning detector like the WX-1000.

WX-950 SERIES

Easy-to-read, the WX-950 has unequaled flexibility and is attractively priced. The WX-950 has a unique strike or cell display mode and precise mapping technology.

WX-1000 SERIES

The WX-1000 is designed for the demands of high-performance aircraft. It can also interface with many GPS/LORANS to simultaneously display weather and navigation data. Moreover, the WX-1000 can be integrated with the SKYWATCH™ Traffic Advisory System.

The WX-1000 Stormscope® is larger and heavier than the WX-900 and has more features, including two views, either 360 degrees or 120 degrees, of forward airspace. The WX-1000+ system features gyro stabilization so weather information is displayed relative to the aircraft's heading, clock display, and the ability to interface with LORAN/GPS receivers and display navigation information on the Stormscope® CRT.

The main advantage of the WX-1000 is that the CRT is easier to read because the CRT's intensity can be adjusted for various ambient light conditions.

Installation Tips for Stormscope® and Strike Finder®

The Stormscope's wiring harness is complex to assemble, from cutting wires and shield preparation, to a detailed need to adhere to careful harness routing. The completed harness is composed of three main subcables, each containing several smaller wires. Other wires are needed for gyro stabilization hookups (for the WX-1000+), and current requirements for the WX-1000 are higher than for the WX-900.

The display unit (CRT) shouldn't be mounted within three inches of electric gyroscopic instruments like artificial horizons and turn coordinators. If so mounted, the effect might be a wobbling or vibrating display on the CRT.

Both the WX-900 and WX-1000 Stormscope® use the same antenna, a design similar to the combined loop/sense antenna used on ADF systems. Three coils of wire wound around a core form the directional heart of the antenna, which is the H-field component of the antenna (see Chapter 7). A plate inside the antenna forms the E-field component.

An important consideration for Stormscope® installations is the distance of the antenna from various noise-producing elements. Following is a list of such elements and the minimum distances these elements should be from the Stormscope® antenna:

1. Cabling systems carrying five or more amperes—28 inches
2. VHF Comm antennas—14 to 24 inches
3. DME/transponder antennas—48 inches
4. ADF antennas—24 inches
5. Strobe light bulbs and power supplies—60 inches
6. Autopilot servos/Trim servos/autopilot amplifiers—36 inches

7. Heating system igniters and fans, air-conditioning fans—60 inches
8. Airborne telephones—48 inches
9. Fluorescent lamp systems—60 inches
10. Access panels—26 inches
11. Control surfaces—36 inches

Troubleshooting Tips

Strikes displayed in error are but one of the potential problems that can occur. These can be the result of onboard avionics, electrical systems, or poorly bonded equipment and surfaces. This type of error can be easy to isolate by simply turning off each aircraft system until the strikes no longer appear.

Bonding is another matter, however, and won't cause consistent symptoms every flight. It is most likely to cause problems during a flight through dust or moisture-laden clouds, or on hazy or humid days. The WX-900 and WX-1000 units are designed with a self-diagnostic capability that provides pilots with error codes that can help Stormscope® repair facilities quickly track down the problem. Some of the problems that are covered by the error codes include wiring problems, onboard equipment radiation, heading inaccuracies, random access memory problems, and antenna errors.

Pilots using Stormscopes® should know that the casual throwing of an electrical switch could cause a splash of strikes on the Stormscope® screen, and so can taxiing over buried electrical cables. On older models, knowledgeable pilots will simply hit the Stormscope's reset button, but with later models, the Stormscope® automatically resets itself.

The most important test to perform on a new Stormscope® installation is the interference or ambient noise test, to ensure that nothing on board the aircraft is affecting the Stormscope's ability to function properly. This test should be performed with all engines running at the highest possible rpm and all aircraft systems functioning.

Because it is possible for other aircraft to radiate their noise and interfere with your Stormscope®, you might have to perform this test well away from other aircraft. Also, there should be no storm activity within 400 miles in any direction.

Following are some of the potential troublemakers:

1. Poor bonding or grounds between equipment and airframe structure.
2. Noise generators injecting RF directly into antenna.
3. Degraded aircraft equipment radiating or inducing noise.
4. Noise on DC line to Stormscope® processor.

Aircraft systems most likely to cause the above problems include:

1. Radios such as COMM, DME, transponder.
2. Motors, generators, alternators, inverters, servos, and actuators.
3. Strobe power supplies (poor design or failure of internal filtering system).
4. Air-conditioning and heater fans and heater igniters.
5. Fluorescent lighting systems.
6. Windshield heat systems.

To begin the test, start the engines and warm them up, then run at full practical rpm with all aircraft equipment on, including the Stormscope®. Once the Stormscope® has initialized, strikes might begin to appear on the display. Reset the system and again observe the display for strikes. If the strikes return, turn the aircraft to a new heading and clear the screen. If the strikes reappear in the same location on the screen, ambient noise is causing the strikes. If the strikes are in a new location, they are probably caused by a storm, if there is one within range.

Other isolated strikes that appear in small amounts can generally be ignored. The rule of thumb is that if the displayed isolated strikes are not clusters of strikes but are sporadically strewn around the screen, they should not present a problem as far as thunderstorm detection or presentation is concerned.

Run through the internal self-test provisions as specified in the user's manual. Covering the entire depth of material on self-tests is not practical because of the varieties of Stormscope® models in the field. Note that COMM radios can cause individual strikes to appear on the Stormscope®. To prevent this problem, the Stormscope® can be inhibited while transmitting on COMM radios. Should two or more microphones be connected to the communications system, isolation diodes must be installed, and a Stormscope® factory-authorized avionics shop should do this job.

WX-900 TROUBLESHOOTING

WX-900 troubleshooting covers three areas: the indicator (display/processor), the antenna, and wiring. If the display/processor does not turn on or show any signs of life, the problem is in either the display unit, internal processor, or the wiring. The first item to check before removing the unit for repair is that it is getting power and is properly grounded. Any mechanic should be able to perform this test. If power and ground check out okay, then the next step is to remove the display/processor and bring it to an authorized Stormscope® repair shop.

WX-1000 TROUBLESHOOTING

Following are some troubleshooting checks for the WX-1000. You can do some of these yourself, but many will require testing by a Stormscope repair shop.

1. If the display presentation does not appear, the green LED will be illuminated. Check for proper power and ground at the processor. Lack of power or a bad ground could be caused by poor seating of the processor in the mounting tray, a popped circuit breaker, broken wires leading to the mating connector, or a defective processor.

2. If the green LED is faintly lit, the processor is getting power and is grounded, but there might not be power or a ground to the PWRSWHI and PWRSWLO connections. This could be caused by broken wires, a defective processor, or the PWR/Brightness switch needs to be switched on.

3. A bright green LED means the processor is getting satisfactory power, however, the display might be defective or the display wiring could be open. It also could be as simple as the display connector being loose.

4. Both green and yellow LEDs lit: This might be an overload to the processor or a system fault. The processor will have to be bench-checked, and the wiring to the display and antenna will have to be examined.

5. Green and red LEDs lit: This indicates the processor has a fault. The processor will have to be bench-checked.

6. Display distorted: The processor and display will have to be bench-checked; in addition, the wiring in the display should be checked. The cause of the problem could be another panel-mounted instrument, such as electrically driven gyroscopic instruments.

7. If the presentation appears to vibrate or wobble, it might be caused by electrical or magnetic interference on the display. Isolate by turning off suspected components until the defect disappears. If this fails to resolve the problem, have the processor and display bench-checked. With the Stormscope® on and other aircraft instruments and systems running, you can slide the display unit out of the panel while it is still hooked up and watch for the wobble to disappear. If it does disappear, this confirms that the interference is coming from something nearby in the panel, most likely a device that projects a magnetic field. You can observe a similar effect by holding a telephone handset or a magnet near a computer screen.

8. Improper brightness control may be observed as “blooming” of the image on the display. This might be a focus problem within the display and will require a factory-authorized repair.
9. Incorrect information displayed on the “system data” screen is probably caused by incorrectly installed option jumpers or a defective processor. This will require bench-checking and factory-authorized repair.

It is important for you to know that troubleshooting recommendations for Stormscopes might require special test equipment, parts, and experience with problems specific to these unique lightning-detecting devices. This chapter is intended to raise some questions, discuss some hints on where the problem might be located, and hopefully help you get on the right track to solving some squawks without emptying your wallet.

What Can Lightning Do?

While getting hit by lightning in a metal airplane will give pause to just about anyone, think about how it badly it would scare the blue blazes out of anyone flying in a composite (“plastic”) airplane without lightning protection. Why? Take a look at how atmospheric physics plays in this drama.

We live on one plate of a really big spherical capacitor. The ionosphere is the other plate of this 1.7-farad capacitor, and the potential difference between these plates is something on the order of a hundred million volts. At any one time in the world there are about 1,500 localized “puncture holes” in the atmospheric dielectric that lets this charge leak from one plate to the other. We call these puncture holes thunderstorms. The leakage current takes the form of a gigantic electrical spark that we lovingly call lightning. Jim Weir of RST Engineering, the only one I know of who effectively received approval to develop and manufacture a Navigation/Communication radio kit, knows his lightning and has this to say:

Anatomy of a Strike

A plain old lightning bolt can sustain a deadly current of about 25,000 amperes during its short, but brilliant life. Let's see now. Ohm's law tells us that power is voltage multiplied by current, so if the math is right, 100 megavolts times 25 kiloamperes comes up a peak power of 2.5 terawatts. That's 25 with 11 zeroes following it. That's a bunch of power.

Damage Report

While the delicates inside the radios didn't like the megavolt pulses, the airframe itself was barely scratched. You see, in a metal airplane, the power dissipated in the airframe is the product of the current (squared) multiplied by the resistance of the airframe from where the lightning bolt enters to where it exits. Even if the airplane were so poorly riveted together that there was a milliohm from one end of the airplane to the other, the power dissipated by the airframe would only be 625 kilowatts. Now that is a lot of power, but it is dissipated all over the metal airframe, and the lightning bolt only lasts for a few milliseconds. The plane may survive, but the avionics will probably not be so lucky.

That's why the thought of a lightning strike in a composite airplane can be so unsettling. You see, composites are a relatively high-resistance structure, actually one big insulator. That's what makes them so wonderful for hidden antennas: The structure is almost transparent to radio waves. But what about lightning? That's another matter. When lightning hits a composite structure ("epoxy"), it carbonizes some of the epoxy, which turns the airplane into a rather large value resistor. This oversized resistor doesn't pass the current through the structure, but rather dissipates the current as heat.

Entrapped within the epoxy is a fair amount of water vapor. When the heat generated from the lightning boils the trapped moisture during a strike, the vapor expands explosively and causes structural disintegration of the epoxy. Not a pleasant thought, especially if your fanny is being held aloft by an airfoil made of disintegrating epoxy and composite fabric.

It seems that, even more than a metal airplane, a composite airplane should be more concerned about a lightning strike. Indeed, the designers of the Learfan destroyed (literally) boxcars full of composite parts during the lightning certification process for that airplane. The only way they ultimately got certification was to weave copper strands into the fiberglass to conduct the current through the copper instead of through the epoxy. Beech used the same basic concept on the Starship.

Something Useful?

Back to the physics books. When any electrical energy consumes itself in an arc (spark), it emits electromagnetic waves that are essentially radio noise. As mentioned earlier, for years we've seen the old ADF needle point at the nearby thunderstorm instead of the distant NDB on the ground. It isn't a particularly difficult phenomenon to illustrate. Next time there is a thunderstorm within 100 miles or so, tune your AM radio to the lowest frequency on the band that doesn't have a station. Every lightning strike will make a characteristic crackle on the radio.

History

In the early 1960s, Jim Weir, a budding engineer (and the source for some of this material) was working for a San Diego consulting firm. He had the job of building receivers that looked for the signature of lightning bolts, so they could distinguish between cloud-cloud and cloud-ground strikes. Along the way, they also learned to do some other interesting stuff like distinguishing between direct strike information and strikes many thousands of miles away that bounced their energy off the ionosphere and back to their receivers in San Diego. The result of this experiment was the discovery that most lightning strike energy was generated between 25 kHz and 2 MHz with a portion going up into the gigahertz range.

Later, in the early 1970s, an engineer named Paul Ryan (no relation to the aircraft company of the same name) used this lightning information to build a lightning detector. However, Ryan went a step further. Instead of just being able to say, "Yup, that was lightning all right," Ryan took the next step and made an ADF-style detector to determine not only the presence, but also the bearing of the lightning.

His next leap of invention was to make the observation that since most lightning bolts have about the same amount of energy, the strength of the signal received at his antenna was relatively proportional to the distance between the lightning and his antenna. Voila! A crude "weather detector" that could tell the distance and bearing to lightning. Refinement built on refinement until Ryan had an instrument that could be used in an airplane for lightning detection and avoidance. Because the usual generator of lightning was a large mass of rain and hail called a thunderstorm, an effective tool was born, that could keep planes and their human cargo from flying in the vicinity of thunderstorms. His system used a narrowband energy detection scheme, which was limiting and not accurate enough.

An engineer named John Youngquist, concerned about the shortcomings of the Ryan method, devised a different approach which looked at a much broader bandwidth that was able to do some signal processing that made the detection more accurate.

Later, Ryan (and his successors) began using more sophisticated digital filtering. Youngquist answered by applying some state-of-the-art computer processing with both methods developing to the point where they are both reasonably reliable lightning detectors.

In the early 1980s, Ryan sold the patent rights and his Stormscope® name to 3M, which in the early 1990s sold the rights to B.F. Goodrich, which today still builds advanced versions of the Stormscope® in Grand Rapids, Michigan.

John Youngquist now owns the patent rights and name Strike Finder®, and his company, Insight Avionics, builds the Strike Finder® (and other avionics devices).

Both systems now play active roles in aviation and will continue to do so for some time. Both are figured into the new array of multiple display systems making their mark on the general aviation scene. Both the Strike Finder® and the Stormscope® are equally efficient in detecting and displaying potentially destructive storm masses. Strike Finder® uses the gas plasma display and Stormscope® uses a CRT, but the price differences are significant. The Strike Finder® prices out for less than the Stormscope®. The numbers can vary depending on options, but even at that, the Stormscope® is still more.

Plastic Planes

Now let's get down to the tips and tricks of installing a lightning detector instrument in a plastic airplane. The techniques and process are basically the same. First, lightning detectors are not for the home brew installers. The amount of test equipment and special instrumentation is prohibitive. In fact, it is not recommended that lightning detectors be installed in "plastic" aircraft. Finding a quiet location is extremely difficult. The following information is for reference only. Installing either the Strike Finder® or Stormscope® into a composite plane is not recommended.

Installation

Skin mapping is a critical step to a successful installation, which simply means moving the antenna around to find the best location where noise presents the least problems to the lightning detector. This is true for both the Stormscope® and the Strike Finder®. To keep the test from being affected by the proximity to the technician; a wood broom handle with the antenna secured at one will work. There will be some "hot" spots and some "cold" spots. The hot spots are surface areas where there is a lot of electrical noise. The goal is to find the coldest spot or least noisy location that will accommodate the antenna. This is where the shop earns its money. For the Strike Finder®, the antenna for the testing is about $1,500, the cable about $10 a foot, and the audio detector and adapter are about $300. At this point there is over $2,000 in test equipment, just for starters. For Stormscope®, the installation test fixture is something on the order of $1,000.

For planning purposes, the antenna for a lightning strike detector has an almost identical footprint to the modern ADF loop antenna. This antenna needs a ground plane roughly a foot in radius from the center of the antenna. This ground plane can be a metal sheet, wire mesh, copper tape, or wire strands embedded in the surface epoxy and attached to the antenna. The antenna ground plane is important, but not as much as finding a good electrically "cold" spot on the skin. Because the antenna, as an inherent part of its design does not have a metal base, the only bond is precipitation static, anti-static coating on the surface of the antenna and extending down and around to the mounting screws is the only protection offered against lightning strikes. Yes, it is very possible for the antenna to be totally destroyed by a lightning strike. For now, technology only allows for a design that essentially acts like an ADF trailing sense antenna, floating off airframe ground.

Be particularly careful not to locate the antenna within four feet of the DME or transponder antennas. There is no choice in this matter. Eliminate, isolate, or resolve the noise areas or do not install the storm detection system. The same is true for both the Stormscope® and the Strike Finder®.

The antenna can go on the inside of a composite skin, and the ground plane material bonded (read glued) to the inside of the skin. However, the only reliable ground plane method is a metal plate extending a minimum of twelve inches in all directions. Composite planes should have a metal weave in the skin to provide adequate protection against precipitation static and lightning. Again, disclaimers—let only a professional install the system, and get an agreement that they will resolve the problems without excessive additional cost to you, the owner. If the shop does not make you aware of the potential for failure in a composite structure then they are not being professional and you should shy away from them. You need to know the truth, not be swayed by the nice toy.

Stormscope/B.F. Goodrich is predominately after the corporate twin and small jet market; besides, the Stormscope® instrument won't fit into a standard three-inch instrument hole; it fits only the airline-style 3ATI hole (the eight-sided squared-off hole). Strike Finder® from Insight, on the other hand, will fit a regular round three-inch instrument hole or the 3ATI hole.

Stormscope uses a CRT (vacuum tube) display, which offers infinite resolution, but as with any CRT the lifespan is limited and more prone to vibration damage. Strike Finder® uses a gas plasma display. The gas plasma theoretically has an infinite lifetime but can only plot 4100 discrete dots on the face of the display. Stormscope uses narrowband in some models and wideband in others, whereas Strike Finder® is a completely broadband system. When choosing between the two for a professional installation on most GA (general aviation) planes, the choice is no longer as clear. Both have two component installation facets and are equally as easy to install.

In any case, only highly skilled professionals such as those that form the core of general aviation aircraft should be installing either of the systems. The Strike Finder® has more options due to panel mounting requirements. It can easily be mounted into a three-inch hole. Also, harness assembly and installation is a little easier.

Troubleshooting

Troubleshooting a storm detection antenna is essentially one of substitution. If an antenna is squawked as not working, replace with another known good unit. A squawk of not painting storms may be a problem from precipitation static. The antenna is coated with an electrostatic paint, but may not meet the specifications required to dissipate the static. Checking for the proper resistance of the paint requires a meg-ohm reading in excess of 5 meg ohms and up to 70 meg ohms. Refer to BF Goodrich's Stormscope® Service Letter SL 148 for further detail. Your FBO can provide this letter for you. An interesting note, the weather detection antenna is the only one not required to directly RF bond to the aircraft skin.

The four ways noise can enter the Stormscope® are:

1. **Radiated into the antenna:** Poor location is the most likely culprit. Try shielding with a metal can or a sheet of aluminum as a troubleshooting technique. Relocation may be necessary.
2. **Radiated into the cabling:** This may be routed with high-current-carrying cables or the shielding at the connectors is inad-

equate. Process of elimination is probably the only way to find the entry point. Try wrapping the portion of the harness near the connector that is not shielded. The shield terminations should extend up to within one inch of the connector. Also, double-check the terminations to make sure they were done correctly.

3. **Riding on the DC power line:** A DC line filter may resolve this problem, but try correcting the source of the noise first. It could be a defective servo, actuator, or alternator. The diodes in the alternator may be defective or the capacitor on the alternator case may be ineffective (replace). Newer and more effective capacitors for the alternator are available from Lonestar.
4. **Different potentials between the antenna and the processor:** Try adding a ground wire between the processor, the antenna ground plane, and to the battery.

If you need DC line or alternator filters for Raytheon or Cessna aircraft, contact their respective customer support departments. Cessna's filter is P/N A263. Lonestar of Texas has an excellent filter that should do the job. The part number is 122253-5A/10A (5- or 10-amp versions).

I wish to acknowledge and thank Jim Weir for his excellent writing skills and insight into elements of this book, especially the sections on lightning. He is known for his definitive positions on areas of avionics and an intimate understanding of his craft.

Thunderstorm Detection System Manufacturers

These manufacturers' addresses can also be found in the Sources section.

B.F. Goodrich Flight Systems (Stormscope®)

Insight Instruments (Strike Finder®).

CHAPTER 20

Compass

If you think you have a problem with your heading indicator, check the accuracy of your compass first. This advice will save you lots of money.

Want to know how to make enemies with an instrument shop? Imagine their frustration when they find nothing wrong with an HSI (horizontal situation indicator) or heading indicator that's been sent in because it precesses or displays the wrong heading. Their frustration builds even higher when they return the unit to the customer, and the customer complains, "It still doesn't work!" Repeatedly, I've seen customers, mechanics, pilots, and kit builders send heading indicators in for repair. The instrument shop goes through the whole exercise and finds the bearing race is worn, possibly enough to cause precessing. Cautiously, for they've been through this many times, they warn the customer, "There is no guarantee this will solve your problem. There is a little bearing recessing, but it shouldn't be enough to cause the degree of error you described. Do you still want to have it repaired?" The customer usually tells them to go ahead, the gyro is repaired, bench-checked, found to meet factory specs, and sent back to the customer. The customer installs it, and the same squawk resurfaces.

The mechanic or owner calls the instrument shop and complains, "It's still doing the same thing." The instrument tech asks if the mechanic checked the indicator on an official compass rose or against a master compass. "Of course we have," is the defensive answer. The "defective" indicator is once again sent back, and the instrument shop finds nothing wrong but spends several hours trying to duplicate the squawk.

Finally, after being pressured by the instrument tech, the mechanic and owner taxi the airplane to a compass rose and check the compass. Sure enough, the compass was way off, and so was the heading indicator, because the pilot set it according to the erroneous compass. One important thing to check is fluid levels in the "whiskey compass." If the fluid level is low, it is considered to be defective. This is not a field repair. Although many mechanics have refilled the compass with the correct fluid, there was a reason it was low and therefore that is the basis for replacement.

What do you suppose the customer will want now? You guessed it; the customer doesn't want to pay for the repairs to the heading indicator. After all, there wasn't anything wrong with it. Nevertheless, the instrument tech did exactly what he or she was asked to do—confirm a defect and repair it. In fact, the tech was instrumental (no pun intended) in determining the cause of the problem and pointing the way to a solution. In many cases, the instrument shop doesn't charge for this kind of repair, but it really is a lot to ask; after all, it wasn't the instrument shop's fault. It is the troubleshooter's fault. The mechanic and/or the pilot should have taken the airplane to an official compass rose and confirmed whether the gyro or the compass was at fault. This isn't a rare occurrence, but sadly, it happens more often than most can afford.

The same scenario happens with the customer trading in a perfectly good heading indicator for an overhauled unit. When the problem isn't fixed by the overhauled gyro, the owner gets upset, and whoever did the faulty troubleshooting is left with labor charges he or she has to swallow.

Compass Construction

The wet compass consists of a sensing elements (bar magnets) attached to a floating card that is free to move inside the compass housing. The compass card essentially stays still, always pointing in the same direction, while the airplane "moves" around the compass. The lubber line on the face of the compass is attached to the airplane and moves with the airplane, thus displaying the magnetic heading of the airplane. To dampen or restrict sudden changes in compass-card motion as the airplane moves around the card, the card is submerged in a fluid with a consistency similar to that of kerosene. Small compensating magnets allow for adjustments to keep the compass accurate.

Vertical Card Compasses

The vertical card compass is a replacement for the time-honored wet compass. The chief advantage of a vertical card compass, which is subject to the exact same errors as a regular compass, is the presentation. It looks and reacts just like a heading indicator, which makes using it much more intuitive.

Vertical card compasses have a massive magnet that is mounted on a shaft supported by jeweled bearings in a vertical housing. As the airplane rotates, an azimuth card connected through gears to the magnet rotates to display the magnetic heading.

Remember that while the presentation is easier to interpret, the vertical card compass is just a compass and must still be calibrated regularly and is subject to normal compass errors.

Compass Problems

The main problem caused by compasses is when people ignore the need to calibrate the compass on a regular basis. After they've spent hundreds of dollars on unnecessary heading indicator overhauls, they learn that it is much cheaper to check the accuracy of the compass whenever they suspect a heading indicator accuracy problem.

Assuming the compass is in good shape physically, here is a reference-only checklist for determining if the compass is the problem when you suspect the heading indicator is inaccurate:

1. Taxi the airplane to an official compass rose and turn on all normally used equipment (radios, lights, etc.). Keep the engine running.
2. With the airplane stationary, record compass deviation when you turn on intermittently used equipment such as electric prop deicer boots, electric fans, air conditioner, etc. You need this information so that you know how far off your compass reads when you're using this equipment.
3. Turn the airplane to north (according to the compass rose) and set the heading indicator to match the compass. Note the reading on the compass, if it isn't exactly north. If the compass is more than 10 degrees off at this point, you need to calibrate it before continuing because the compass is out of legal limits.
4. Turn to east according to the heading indicator and record the magnetic compass reading.
5. Turn to south according to the heading indicator and record the magnetic compass reading.
6. Turn to west according to the heading indicator and record the magnetic compass reading.
7. Turn back to north according to the heading indicator. The magnetic compass should be within 3 degrees of what it read in step 3.

If the magnetic compass and the heading indicator match in step 7, the heading indicator is fine. Any errors more than 3 degrees (difference between the compass and the heading indicator) on the other cardinal headings are caused by the compass, and the compass should be swung.

If when you turn back to north according to the heading indicator and the compass is more than 3 degrees different than north, you need to line back up on the north direction on the compass rose. After doing so, verify that the compass is reading the same as when you lined up on north in step 3. Note the heading indicator reading. If it is more than 3 degrees off from the compass reading, it is probably precessing too much. FAA standards call for no more than 3 degrees precession every 15 minutes, and if you've completed this exercise in less than 15 minutes and the heading indicator is more than 3 degrees off, then the gyro is due for some attention.

Compass Location

Each aircraft has its own unique magnetic pattern, like a fingerprint. The structure, control yoke, radios, engines, props, landing gear, and hardware all create magnetic influences, and these must be compensated for in order to obtain accurate information from your compass.

The factory location for your compass is usually the best place to put it. After all, the factory has determined that it's possible to obtain the necessary accuracy from that compass location. But over time, some things can change that will affect the compass and throw it off more than the allowable 10 degrees.

One problem that can happen on older steel-tube and fabric airplanes is that the steel fuselage gets magnetized over time. There is no way to prevent this; it's just a consequence of steel existing in a magnetic field (the earth's). If you had a big enough degausser (demagnetizer), you could theoretically demagnetize the airframe by running the airplane through the degausser. Attempts to demagnetize airframes using a portable degausser aren't so effective, because the magnetized area disappears where degaussed but then shows up somewhere else on the airframe. So what do you do if your airplane is hopelessly magnetized and you can't swing your compass to within 10 degrees?

The only option is to either install a remote compass whose sensor is mounted out on a wing away from the fuselage, or move the compass to an area where it is more accurate. We had this experience on a Citabria once. No matter how hard we tried; we couldn't get the compass swung within 10 degrees. The compass, by the way, was freshly overhauled. Finally, one of our brilliant mechanics came up with the idea of repositioning the compass. He experimented with various locations and was able to get great accuracy by moving the compass, normally mounted directly in front of the pilot on the top center of the glareshield, a few inches to the left of its original location. Problem solved, customer happy.

Compass Repairs

Unfortunately, compasses are another item that the pilot isn't allowed to work on. In fact, you aren't even permitted to swing your compass, unless under the supervision of a licensed mechanic. Please note that licensed A&P mechanics are not permitted to overhaul compasses because a compass is an instrument, and instrument overhauls are major repairs that can be done only by certificated instrument shops or by a licensed repairman. Sure, all the discount suppliers sell compass repair kits and compass fluid, but if you want legal components in your airplane, an authorized shop must do the compass overhaul. Compass removal and installation, of course, must be done by a mechanic or under a mechanic's supervision.

If your compass is starting to look yellow and the front glass is hard to see through, you might consider getting it overhauled, even if is still accurate. When you see bubbles in the fluid or when half the fluid is gone, it's not worth simply topping up the fluid because it's leaking from somewhere and the leak needs to be fixed. Over time, the gaskets that hold in the fluid get brittle and crack, and when you smell the kerosene odor of compass fluid, those worn gaskets are where it's coming from. It's time for an overhaul. After the overhaul or any installation of new equipment, the compass should be swung.

Don't forget that FARs require a working compass for any type of flying, so proper maintenance of the compass is essential. A nonworking or inaccurate compass is a no-go item.

Compass Swinging

To swing the compass, you'll need to find an official compass rose. Don't use a compass rose that's been painted on the ground outside the maintenance shop. Compass roses have to be carefully designed and located in areas free of magnetic interference. Other criteria include using nonmetallic paint to draw the lines, proper radius for the size of aircraft using the pad, weight-bearing ability of the pad for use by large aircraft, and freedom from intermittent local disturbances such as nearby factories that use high current for short durations. If you're not sure of the quality of a local compass rose, ask the local FAA inspectors if the rose meets FAA Advisory Circular 150/5300-13, Appendix 4 specifications.

Next, you'll need a nonmagnetic nonferrous flathead screwdriver for adjusting the compass. You can make one easily by filing the tip of a piece of stainless steel welding rod. Obviously, using a ferrous or magnetic screwdriver would make it impossible to adjust the compass's compensating magnets accurately, because each time you brought the screwdriver near the compass, the compass card would start spinning wildly. You might have seen this happen when you put your headphones on the dashboard next to the compass. In fact, it's a good idea to keep headphones away from compasses because the magnets in the headphones can permanently damage them.

The flux valve (or flux gate) remote compass is mounted with three nonmagnetic screws to a mounting plate with three locknut assemblies. The valve is engraved or marked to indicate "FWD," because the valve will only work correctly if properly aligned. Reference marks next to the "FWD" symbol are used during calibration to set the valve to correctly display magnetic headings on the compass card.

Compass swinging is easier with two people—one to taxi the airplane and adjust the compass, and the other to stand outside and direct the airplane to the proper spot on the compass rose. It's hard to assess alignment of the airplane with the compass rose from inside the cockpit, unless the compass rose has detents against which you can stop your wheels when you are lined up. You can make the job easier, too, if the outside person has a hand-held transceiver.

Here's the step-by-step compass swinging procedure:

1. Taxi the airplane to the compass rose and line up so the airplane is pointing to magnetic north according to the compass rose. Make sure all radios and normal equipment are on, such as flashing beacon, strobes, and landing light (if you are in the habit of flying with it on all the time). Keep the engine running fast enough so that the generator stays on line. If it can be done safely, you might want to run the engine near the low end of the cruise range, usually around 1800 or 1900 rpm to more closely replicate actual flying conditions. Note that if you are swinging the compass on a taildragger, you'll need to rig up some way to keep the airplane in a level attitude while you swing the compass. An alternative method would be to swing the compass against a calibrated master compass while flying. Ask your avionics shop if it has a master compass.
2. Adjust the north-south screw on the compass so the compass card reads exactly north.
3. Taxi around and line the airplane up with east on the compass rose. You won't be able to simply turn right 90 degrees; you'll have to make a wide left 270 degrees to get lined up on the compass rose properly.
4. Adjust the east-west screw on the compass so the compass card reads exactly east.
5. Taxi back around and line the airplane up with south on the compass rose.
6. Note the amount of deviation from south as shown on the compass. The aircraft is lined up with magnetic south according to

the compass rose, but the compass will show some heading other than south, like 170, or 185, or 165.

7. Using the north-south adjusting screw on the compass, carefully remove half the deviation shown on the compass. If the compass reads 170 before the adjustment, turn the north-south screw so that the compass reads 175. The deviation is 10 degrees, and removing half the deviation brings the compass card 5 degrees closer to 180. If the compass reads 165, adjust the screw so the compass settles on 172.5 degrees.
8. Taxi to line up with west on the compass rose.
9. Note the deviation from west shown on the compass and remove half of it using the east-west adjustment screw.
10. Now you're ready to check the accuracy of the compass. Taxi back to north on the compass rose and note the deviation. It should be about the same as the deviation you obtained after adjusting the compass for south. If the south reading ended up being 5 degrees off, then you should see a 5-degree deviation on north. Remember that you're shooting for anything less than 10 degrees deviation on any heading. By making these four cardinal adjustments first, if they are below 10 degrees, then the compass should be within 10 degrees on all headings. Taxi to east and check the deviation there. Again, it should be about the same as the west deviation (after adjustment). If the north and east deviations are very different from south and west, you might need to make some careful, slight adjustments to the adjustment screws, but if the deviations are close, leave the adjustments alone.
11. On a piece of paper, note the twelve compass headings that are listed on the compass deviation card: north, 30, 60, east, 120, 150, south, 210, 240, west, 300, 330.
12. Starting with north, note the compass deviation at each point, with the airplane lined up on the proper magnetic heading according to the compass rose. Note the deviation either as a plus or minus from what the compass should read (+3, –2) or as the heading shown by the compass for a particular compass

rose heading (92, 154, etc.), whatever suits your methods. I prefer the plus or minus method because you add or subtract the deviation to your magnetic heading to obtain compass heading.

13. Fill out the compass correction card, note the date and the conditions (radios on, lights on, etc.), and install the compass card in the airplane.
14. Ask the mechanic or other qualified person who helped you swing the compass to make an airframe logbook entry and sign it. Make sure the logbook entry includes the deviations found during the swinging. That way, if you lose the compass correction card in the airplane, you won't have to swing the compass again. You can just make a new card using the numbers you so wisely entered in the logbook.

Slaved Compass

Slaved compasses usually aren't affected by additional equipment because their sensing units are mounted well away from the cockpit in one of the wings. They should be checked for accuracy whenever you swing the compass. If inaccurate, the remote compass's flux valve or compass amplifier might need to be adjusted by your avionics shop.

Two frequent mistakes made with remote compasses are using magnetic screwdrivers to unscrew them from their mounting points in the wing, and reinstalling the compass backwards after a routine inspection. On Piper Seminoles, for instance, the remote compass is mounted to a right-wing inspection panel. A careless mechanic can easily remount the inspection panel 180 degrees off (the panel installs either way and isn't marked), and you'll only find out when you notice the compass card on the HSI is 180 degrees off next time you fly. If you have a remote compass installed in such a fashion, have an arrow and legend applied to the panel so it's obvious which way the panel mounts to the wing. The panel should also be placarded: "Remote Compass: Do Not Use Magnetic Tools."

One word of caution: Intermittently used equipment like windshield heater, prop deicer, and heater fans can affect the compass. You should try each item separately and together and note how far off the compass is when these items are switched on. You don't want to calibrate the compass with these items on because they aren't used for normal flying, but you should be aware of how far off they will throw your compass so that you can compensate when using the equipment in IFR conditions.

Compass Manufacturers

These compass manufacturers' addresses are listed in the Sources section of this book.

Airpath

Hamilton Instruments (vertical card)

Instruments & Flight Research

Precision Aviation (vertical card)

CHAPTER 21

Gyroscopic Instruments

Gyroscopic instruments include turn coordinators and heading, attitude, and horizontal situation indicators.

Heading Indicator

The heading indicator, or directional gyro, is driven either electrically or by air blowing across a set of scoop-like paddles. The vertical compass card display that is seen by the pilot is driven through gears from the gimbal. Just as a toy gyroscope or top is resistant to movement once it reaches top speed, so is the gyroscope in the heading indicator. When the aircraft changes direction, the gyro maintains its position and the aircraft "turns" around the stationary gyro. Thus, the instrument's compass card shows that the aircraft is on a new heading. The jeweled bearings that compose the supporting mechanism for the gyro are very sensitive to any form of contamination such as dust and moisture, but if the air that drives the instrument is kept clean and at the correct flow rate, the heading indicator is a reliable and dependable instrument.

The primary problem with gyroscopic instruments is their tendency to precess. You'll notice this as a tendency of the heading indicator to drift off a chosen heading over a period of time. FAA specs call for no more than 3 degrees of precession, each 15 minutes, and even with normal precession, you'll have to reset the heading indicator to match your compass every 15 minutes. If you have a slaved heading indicator or HSI, the indicator's compass card will automatically be kept lined up with the compass; your job becomes simply monitoring the heading indicator to make sure it matches the compass. Precession on air-driven gyros is caused by dirty filters, vacuum pump failure, or worn or damaged gyro bearings. Exceeding operational limits of the gyro such as degrees of pitch or bank can cause the heading indicator to spin crazily. This can also be caused by an internal failure of the gyro.

Attitude Indicator

In order to fly IFR safely, pilots must have some way to discern the attitude of their aircraft. Without the attitude indicator and its ability to show the relationship between the aircraft and the horizon, the pilot is virtually cut off from the outside world.

According to Dr. Richard Jensen's book *Aviation Psychology,* a study of military aviators found that except during straight-and-level flight, the highest percentage of pilot eye-fixation time was devoted to the attitude indicator. Although the airplane can be flown IFR without the attitude indicator, this instrument is essential to safe IFR operations.

The attitude indicator isn't too complex. It is essentially a gyroscope that is spun by airflow or electric power. The gyro spins in a double gimbal with the spin axis in a vertical plane. The "airplane" symbol on the face of the instrument is attached directly to the instrument, which is solidly fixed to the airframe. Thus, the "airplane" symbol always does exactly what the airplane is doing, turning, climbing, etc. As the airplane changes attitude, the gyroscopic horizon remains parallel to the earth's horizon, and the pilot can discern the relative motion of the airplane by comparing the instrument's symbolic airplane with the fixed "artificial" horizon.

Information about the aircraft's attitude and changes in its attitude can be output to autopilots via potentiometers, contact points, LEDs, or even lasers. The autopilot uses these signals to make corrections to the aircraft's attitude as commanded by the pilot through the autopilot controls.

Modern attitude indicators have wide limits for safe operation, usually a maximum of 70 degrees pitch and 100 degrees of bank. Pushing the instrument beyond these limits can upset the gyro, resulting in the horizon bar swinging back and forth. Flying straight-and-level should eventually settle the gyro down. If your instrument has manual caging, engage this to settle the gyro down immediately.

Attitude Indicator Installation

Attitude indicators must be installed so that the instrument is accurate in a particular aircraft. Some aircraft panels, for instance, are not perfectly vertical, and the attitude indicator must be adjusted to reflect this. Adjustments to compensate for the tilt of the aircraft's panel are made by adjusting the instrument's rotor housing to match the earth's horizon. The instrument shop mounts the indicator to a tilt table set to the same angle as the aircraft's panel and aligns the rotor housing and pivot assembly for proper indication. If the tilt is too aggressive, a tapered shim might have to be installed.

Small roll adjustments are made on the airplane; most attitude indicator mounting holes are slotted to allow minor adjustments when installing the instrument. To get the setting correct, you might have to fly the airplane and have a mechanic make the adjustment while you hold the wings level. On the other hand, simply level the airplane on the ground using a long carpenter's level set on the control columns to show when the airplane is level on the ground.

Most air-driven attitude indicators don't have a warning flag to alert the pilot to incorrect indications, however, one manufacturer, Sigma-Tek, produces an attitude indicator with a warning flag. A warning flag is a good idea because it can be difficult to determine that the attitude indicator is failing (if you don't notice that the vacuum gauge reads zero). The indicator will start tilting gradually as it spins down. If it isn't evident that the vacuum pump has failed, you might initially try to follow the tilting attitude indicator until you notice that the other instruments aren't backing up what the attitude gyro is telling you.

Turn Coordinator

Turn-and-bank instruments and turn coordinators have a gyro suspended in the longitudinal axis that spins in alignment with the airplane's lateral axis. Most modern turn coordinators are electric-powered so as to provide a backup gyro instrument in case of vacuum-pump failure. Some notable accidents have occurred that could have been caused by the pilot's not realizing that all three gyro instruments, including the turn-and-bank, were air-driven. When the vacuum system failed, so did the only instruments that could help the pilot keep the airplane upright.

Electric turn coordinators have a warning flag that lets you know if the instrument loses its electric power.

Horizontal Situation Indicator

The horizontal situation indicator (HSI) combines two instruments: the heading indicator, and the CDI with a glideslope needle. This makes the pilot's workload much easier and flying using instruments safer.

The typical HSI contains the following items:

1. Compass card.
2. Heading flag.
3. Autopilot heading bug.
4. Slaving meter.
5. Lubber line.
6. Course arrow.

7. NAV flag.
8. Miniature airplane.
9. Course reciprocal pointer.

Gyro Maintenance

Gyroscopic instruments don't require much in the way of aggressive maintenance beyond routine filter changes and gyro overhauls. Outright gyro failures caused by internal wear usually occur after one thousand or more hours of gyro operation. If you're looking for a suitable time period for scheduling gyro overhauls so as to prevent failures rather than wait for them to happen, 1,000 hours would be a good figure. If you don't fly that much and it takes years to accumulate that much time, every five years is a good interval for gyro overhauls and shouldn't add too much to your operating costs.

Bearing-to-race friction wear, inner or outer gimbal surface erosion, and heat all play a major part in internal component failure. Because proper gyro operation depends on minimum friction and an adequate flow of air, it doesn't take much to reduce efficiency. Minute particles of dust, tobacco smoke, and other contaminants can clog filters and reduce efficiency. Gyros are lasting longer these days, possibly because of the increased emphasis on the hazards of smoking, especially for pilots, but also because of improved cockpit housekeeping and careful handling of delicate gyros because of their high cost.

The easiest way to keep gyros healthy is regular filter changes. Most aircraft use a vacuum pump to suck air through the gyros. Before the air gets to the gyros, it goes through a large filter called the central vacuum filter. This is a fine filter that catches most particles and keeps them from getting into the gyros. You can tell if this filter is clogged by checking its color. Anything other than a clean, white appearance means the filter is ready to be changed. In an airplane owned by a smoker who smokes while flying, you'll see this filter turn yellow from the cigarette tar. The more clogged this filter, the harder your vacuum pump will have to work to provide the suction necessary to drive the gyros, and the shorter the life of the vacuum pump.

The central vacuum filter should be changed at least every 500 hours or once a year if the airplane doesn't fly that much. Vacuum pump manufacturer Airborne recommends changing the central vacuum filter whenever installing a new vacuum pump. But if you change the filter on a regular basis as suggested above, it might not be necessary to change the filter when changing the pump. Airborne's advice is probably due to the fact that many mechanics ignore the central vacuum filter and change it only when it looks dirty. Many aircraft, like Piper's Navajo, use enclosed canister type central vacuum filters, and you can't evaluate the filter's condition by looking at it because the filter is inside a can.

I worked on an airplane once where the suction was zero even though the vacuum pump was working. The central vacuum filter was so clogged that the pump couldn't suck any air through it. A new filter brought the system back to normal, although the vacuum pump had to be changed shortly thereafter because it had worked so hard trying to suck air through the clogged filter that it wore itself out prematurely.

The other filter in the vacuum system is the vacuum regulator filter. This is installed at the vacuum system regulator and lets air into the system just before it goes through the vacuum pump to regulate the level of suction in the vacuum system. This filter isn't as critical to the instruments, but does prevent dirt from getting to the vacuum pump, so it is important for vacuum pump life. Replacement of this filter should be done every one hundred hours, and the filter is inexpensive enough that it is worthwhile to replace frequently.

The rubber or vinyl hoses that direct air to the gyros are important to the safety of the plane. Kinked or twisted hoses could cause sluggish gyro operation, and these should be inspected regularly and carefully, especially at the annual inspection. Excessive heat is a factor that can shorten gyro life. Expensive damage can happen to gyros in aircraft that sit outside without any protection from the sun. On a bright sunny day, the sun's energy passing through the windows creates a greenhouse effect, trapping hot air inside the cockpit. With each passing moment, heat levels rise, placing tremendous stress on everything inside the cockpit. This includes the electronic components in avionics equipment, nylon mounts, plastic housings, and even silicon lubrication inside indicators.

Problems caused by heat include tolerance variations from thermal expansion, resistance changes in electronic components, and thinning of lubricants. After the aircraft has been baking for an extended time, the pilot jumps into the cockpit, throws open the windows and doors, and tries to cool the cockpit down. After starting the engine, the vanes in the gyros start to spin, but friction on the gyro's bearings is higher than normal. The result is a huge influx of gyro overhauls during the hot summer months.

The jeweled movements (bearings) used in gyros are made from artificial rubies and are extremely sensitive to shock and vibration. They can shatter if bumped hard enough. Gyroscopic instruments should be handled delicately. In fact, all instruments should be handled with great care. Some electrically powered instruments have jeweled movements, while engine instruments usually have simple brass bearing surfaces that aren't as sensitive. To be safe, handle all instruments as if they are more fragile than eggs. An egg can be dropped a quarter of an inch without damage. A gyro dropped the same height—which can impose up to a 20-G load on bearings—won't show damage initially, but the damage will be evident later when you have to overhaul the unit prematurely.

Pilots and technicians should be aware of the signs hinting at gyro problems, including excessive precessing, slowness to erect, spinning, or tumbling. Precessing is normally evident when the gyro spins up initially, and spinning or tumbling can occur during shutdown as the gyro slows down. Tumbling is an indication that the gyro is beginning to wear out and should be checked and repaired immediately. Spin-up usually takes about 90 seconds and spin-down about 10 minutes. Be careful not to bump or bang the airplane while the gyros are spinning down because this can damage gyro bearings. After shutting down the engine, don't move the airplane for about 15 minutes. Even then, move the plane gently to avoid damaging gyros. Avionics systems should not be shut down until the airplane is at a complete standstill. Otherwise, electric gyros will start to tumble and bang around, possibly causing expensive damage.

Do not tap on gyros during the shutdown phase, or any time while they are running, because this could conceivably cause flat spots on the bearings. Proper handling of gyros during removal and installation is also very important. For long life, gyros should be stored on soft materials foam designed to absorb the environmental G-forces (vibration caused by passing trucks, heavy feet, etc.), which can shorten the lives of stored gyros.

When moving a gyro from the shop to the aircraft, the gyro should rest on a foam pad and not simply be sitting on a hard cart or toolbox. If you see someone mishandling a gyro intended for your airplane, speak up and either refuse to have that gyro installed, or ask for a one thousand-hour extended warranty. Remember, as the pilot, you are responsible for your plane, so don't compromise its safety or yours.

Gyro Troubleshooting

Use this checklist to help isolate gyro problems in the attitude indicator.

1. Gyro shaking and spin-up during start should not exceed 1.5 minutes, assuming vacuum is set to proper level.
2. Gyro visual precessing (tilting away from level) during shutdown should last 5 or more minutes, and the gyro should still be spinning after the display has settled, for a total of 10 minutes after shutdown. Less than 5 minutes precessing and 10 minutes gyro spinning mean the bearings are worn, causing increased friction and a quicker shutdown.
3. During shutdown, the index pointer will swing partially across the horizon, dropping both right and left as it settles to the static position. This is perfectly normal.
4. In flight, the horizon should move smoothly in bank and pitch changes.

5. Continued wild shaking or tumbling of the gyro are signs of impending failure.
6. The wing-level pointer should not scrape against the moving parts of the instrument while the aircraft is pitching or banking.
7. Listen for unusual sounds such as a high-pitched whine, whistle, rumbling, or grinding noises. Grinding and rumbling indicate bad bearings. The high-pitched whine means the rotor is turning too fast, probably because the vacuum is set too high. A whistle indicates an air leak. A common place for this sound is around the shaft of the adjuster knob with the airplane symbol. Putting too much side pressure on the adjuster knob shaft can break the seal around the shaft, causing an air leak and the subsequent annoying whistling noise. The leak usually isn't enough to affect gyro operation, but it should be fixed because it can let contaminated air directly into the instrument, bypassing the filters.
8. If the autopilot is not following the gyro output, the pitch or roll pick-offs could be open or the contacts could be dirty.

Gyroscopic Instrument Manufacturers

See the Sources section for the addresses of these manufacturers.

Instruments & Flight Research

RC Allen

Edo-Aire

HSI/Slaved Compass System Manufacturers

See the Sources section for the addresses of these manufacturers.

Bendix/King

Century Flight Systems

Collins Avionics

S-TEC

Sigma-Tek

CHAPTER 22

Radio (Radar) Altimeters

Radio altimeters are frequently and incorrectly called radar altimeters. Radio altimeter transmissions aren't pulsed like radar signals. They are continuous, constant-amplitude frequency-modulated (FM) carrier of 4300 MHz. The unit's transmitter/receiver computes the difference in frequency of the transmitted signal compared to the received signal. This difference is converted to height above ground. A knob on the front of the radio altimeter is used to select the decision height and to test the unit. When the aircraft reaches the selected decision height, the decision height (DH) light illuminates and, with some systems, a horn sounds. Operating altitude is intended to operate no higher than twenty-five hundred feet above the ground with the primary purpose being final approach.

There are pitch and bank limits beyond which radio altimeter accuracy falls off. If you have a problem with your radio altimeter, make sure to note the altitude at the time of failure, weather conditions, and aircraft attitude. Too abrupt an angle to the ground (max 30-degree bank and 20-degree pitch) and logically, the system would have difficulty reading altitude.

Installation Tips

When having a radio altimeter installed, finding enough real estate to mount the antenna can be a problem. The preferred spot is an area located on the belly of the aircraft where a 120-degree cone is clear of obstructions such as other antennas, gear doors, spoilers, and other protuberances.

The transmitter/receiver mounts on the bottom inside the fuselage skin near the antenna. Make sure the installer mounts the antenna using nutplates so it can be removed from the outside, without requiring someone inside the airplane holding nuts while the other person unscrews the antenna-mounting hardware. Besides, getting to the unit can very intimidating and expensive to access.

Two antenna types are generally available: one for horizontal mounting on a flat belly, and the other for mounting on "uphill" locations like the aft fuselage. The uphill antenna is internally skewed to compensate for the upswept angle of the tailcone. Proper antenna bonding is important for radio altimeter antennas.

Coaxial antenna cable should have sufficient room to prevent bends with a radius of less than 3 inches. Otherwise, reflected power can exceed specifications, resulting in reduced sensitivity. There isn't much else that can be said about the system, except to treat it like any transmitter/receiver when troubleshooting, just a bit more sensitive to location, routing, and mounting.

Radio Altimeter Manufacturers

These manufacturers' addresses are listed in the Sources section of this book.

Bendix/King

Collins Avionics

Honeywell

Sources

Suppliers

ACK Technology
440 West Julian Street
San Jose, CA 95110
(408) 287-8021

Acousticom
Champlin Industrial Park
28180 Clay Street
Elkhart, IN 46517
(219) 293-0534
Fax: (219) 294-7250

Aero Mechanism
20960 Knapp Street
Chatsworth, CA 91311-6161
(818) 709-2851

Aerosonic
1212 North Hercules Ave.
Clearwater, FL 34625
(813) 461-3000
Fax: (813) 447-5926

Aire-Sciences
216 Passaic Avenue
Fairfield, NJ 07006
(201) 228-1880

Airpath
13150 Taussig Avenue
Bridgeton, MO 63044
(314) 739-8117

Ameriking
20902 Brookhurst Street,
Unit 107
Huntington Beach, CA 92646
(714) 963-6977

Arnav Systems
22007 Meridian East
Graham, WA 98338
(206) 847-3550

Artex Aircraft Supplies
P.O. Box 1270
Canby, OR 97013
(503) 266-3959

Audio COMM
395 Freeport #21
Sparks, NV 89431
(702) 331-2992

Aviall
7555 Lemmon Avenue
Dallas, TX 75209-0086
(214) 357-1811

Aviation Communications
1025 W. San Bernardino Rd.
Covina, CA 91722
(818) 967-4183
Fax: (818) 332-7563

Becker Avionics
6100 Channingway Blvd
Suite 303
Columbus, OH 43232
(800) 962-2094

Bendix/King Honeywell
G.A. Avionics
400 North Rogers Road
Olathe, KS 66062-1212
(913) 768-3000
Fax: (913) 791-1302

B.F. Goodrich Flight Sys.
2001 Polaris Parkway
Columbus, OH 43240-2001
(614) 825-2002
(800) 544-5759

Bose
The Mountain
Framingham, MA 01701-9168
(508) 879-7330 ext. 4932
(800) 242-9008
Fax: (508) 872-8928

Brittain Industries
3266 North Sheridan Road
Tulsa, OK 74115
(918) 836-7701
Fax: (918) 836-7703

BVR Aero Precision
5459 Eleventh Street
Rockford, IL 61109
(815) 874-2471
Fax: (815) 874-4415

Century Flight Systems
P.O. Box 610
Municipal Airport
Mineral Wells, TX 76067
(817) 325-2517
Fax: (817) 325-2546

Collins Avionics
Rockwell International
400 Collins Road NE
Cedar Rapids, IA 52498
(319) 395-4085

Communications Specialists
426 West Taft Avenue
Orange, CA 92665-4296
(714) 998-3021
(800) 854-0547
Fax: (714) 974-3420

Comtronics
62 County J
Almond, Wl 54909
(715) 366-7093

Concept Industries
2651 Pacific Park Drive
Whittier, CA 90601
(310) 699-0918

David Clark
360 Franklin Street
Box 15054
Worcester, MA 01615-0054
(508) 756-6216
Fax: (508) 753-5827

Dorne and Margolin
2950 Veterans Memorial Hwy
Bohemia, NY 11716
(516) 585-4000

Electro-Voice
600 Cecil Street
Buchanan, MI 49107
(616) 695-6831

11 Morrow
(no longer builds aircraft radios)
P.O. Box 13549
Salem, OR 97309
(503) 581-8101
(800) 742-0077

Emergency Beacon
15 River Street
P.O. Box 179
New Rochelle, NY 10801
(914) 235-9400

Evolution
18097 Edison Avenue
Chesterfield, MO 63005
(800) 859-9550

Flightcom
7340 SW Durham Road
Portland, OR 97224
(503) 684-8229
(800) 432-4342
Fax: (503) 620-2943

Flight Technology International
1571 Airport Road
Charlottesville, VA 22901
(804) 978-4359

Garmin
1200 E. 151 Street
Olathe, KS 66062
(913) 397-8200
Fax: (913) 397-8282
http://www.garmin.com

Hamilton Instruments
106 Neuhauf Street
Houston, TX 77061
(713) 644-0923

Honeywell Business & Commuter Division
5353 West Bell Road
Glendale, AZ 85308
(602) 436-8000

Icom America
2380 116th Avenue NE
Bellevue, WA 98004
(206) 454-7619

Insight Instruments
Box 194 Ellicott Station
Buffalo, NY 14205-0194
(716) 852-3217
(416) 871-0733

Instruments & Flight Research
2716 George Washington Blvd
Wichita, KS 67210-1585
(316) 684-5177
(800) 373-7627
Fax: (316) 684-0140

Kollsman
4729 Palisade
Wichita, KS 67217
(800) 558-5667

McCoy Avionics
10761 Watkins Road
Marysville, OH 43040-9544
(513) 642-8080
(800) 654-8124
Fax: (513) 642-0220

Meggitt Avionics
10 Ammon Drive
Manchester, NH 03103
(603) 669-0940
Fax: (603) 669-0931

MicroCom
Division of Microflight Products
16141-6 Pine Ridge Road
Fort Myers, FL 33908
(813) 454-6464
Fax: (813) 454-6652

Mid-Continent Instrument
7706 East Osie
Wichita, KS 67207
(316) 683-5619
(800) 821-1212
Fax: (316) 683-1861

Narco Avionics, Inc.
(builds replacement radios for Narco)
270 Commerce Drive
Fort Washington, PA 19034
(800) 223-3636
(215) 643-2905
Fax: (215) 643-0197

Oregon Aero
P.O. Box 5984
Aloha, OR 97006
(503) 649-4778
(800) 824-5978

Peltor
63 Commercial Way
East Providence, RI 02914
(800) 327-6833

Pilot Avionics
24212 Solonica Street
Mission Viejo, CA 92691
(714) 474-0401

Plantronics
345 Encinal Street
Santa Cruz, CA 95060
(408) 458-4481

Pointer
1027 North Stadem Drive
Tempe, AZ 85281
(602) 966-1674

Precision Aviation
8124 Lockheed Street
Houston, TX 77061
(713) 644-7383

PS Engineering
9800 Martel Road
Lenoir City, TN 37771
(615) 988-9800
Fax: (615) 988-6619

Puritan-Bennett Aero Systems
111 Penn Street
El Segundo, CA 90245
(213) 772-1421

Ryan International
4800 Evanswood Drive
Columbus, OH 43229-6207
(614) 885-3303
(800) 877-0048
Fax: (614) 885-8307

Senheiser Electronics
6 Vista Drive
Old Lyme, CT 06371
(203) 434-9190

Sigma-Tek (& Edo Aire)
1001 Industrial Road
Augusta, KS 67010-9566
(316) 775-6373
Fax: (316) 775-1416

Sigtronics
822 North Dodsworth Ave.
Covina, CA 91724
(818) 915-1993

Softcomm
2651 Pacific Park Drive
Whittier, CA 90601
(213) 699-0918
(800) 255-2666
Fax: (213) 692-8947

Sony
Sony Drive
Park Ridge, NJ 07656
(201) 930-1000
Fax: (201) 573-8608

Sporty's Pilot Shop
Clermont County Airport
Batavia, OH 45103
(513) 732-2593
(800) 543-8633

S-TEC
(now owned by Meggitt Avionics, Inc.)
946 Pegram
Mineral Wells, TX 76067-9594
(817) 325-9406
Fax: (817) 325-3904

Telex Communications
9600 Aldrich Avenue
S. Minneapolis, MN 55420
(612) 884-4051

TKM Michel
14811 North 73rd Street
Scottsdale, AZ 85260
(602) 991-5351
(800) 233-4183

Trimble Navigation
(No longer produces aviation radios)
Avionics Division
2105 Donley Drive
Austin, TX 78758
(512) 873-9100
(800) 767-8628
Fax: (512) 345-9509

TransCal Industries
16141 Cohasset Street
Van Nuys, CA 91406
(800) 423-2913

United Instrument
3625 Comotara Avenue
Wichita, KS 67226
(316) 636-9203
Fax: (316) 636-9243

Val Avionics
3280 25th Street SE
P.O. Box 13025
Salem, OR 97309-1025
(503) 370-9429
Fax: (503) 370-9885

Wag Aero
P.O. Box 181
1216 North Road
Lyons, WI 53148
(414) 763-9586
(800) 766-1216
Fax: (414) 763-7595

Check out the following vendors and manufacturers for the latest in avionics tools:

DMC Daniels
P.O. Box 593972
Orlando, FL 32859-3872
526 Thorpe Road
Orlando, FL 32824-8133
(407) 855-6161
Fax: (407) 855-6884

EDMO
Felts Field Airport
5505 E. Rutter Avenue
Spokane, WA 99212

Van Nuys Airport
Van Nuys, CA 91355

Meacham Field Airport
Fort Worth, TX 76106
http://www.edmo.com

Jensen Tools Inc.
7815 S. 46th Street
Phoenix, AZ 85044
(602) 453-3169
(800) 426-1194
Fax: (602) 438-1690
Fax: (800) 366-9662
http://www.jensentools.com

The following distributors can help you find a local dealer for your avionics needs:

Avionics International Supply
1750 Westcourt Road
Denton, TX 76207
(800) 553-2233
Fax: (940) 566-8656

Dallas Avionics
2525 Santa Anna Avenue
Dallas, TX 75228-1671
(800) 527-2581
Fax: (214) 320-1057

Hawkins Associates
Company, Inc.
P.O. Box 430
Argyle, TX 76226
(800) 433-2612
Fax (940) 240-0123
http://www.hawkinsassoc.com

Magazines

Aviation Consumer Magazine
75 Holly Hill Lane
Greenwich, CT 06836-2626
(203) 661-6111
Fax: (203) 661-4802

Avionics Magazine
7811 Montrose Road
Potomac, MD 20854
(301) 340-2100

Avionics News
P.O. Box 1963
Independence, MO 64055
(816) 373-6565

Avionics Review
75 Holly Hill Lane
Greenwich, CT 06836-2626
(203) 661-6111
Fax: (203) 661-4802

Index